Technology

The focus of this series is engineering, broadly construed. It covers technological innovation from a range of periods and cultures, but centres on the technological achievements of the industrial era in the West, particularly in the nineteenth century, as understood by their contemporaries. Infrastructure is one major focus, covering the building of railways and canals, bridges and tunnels, land drainage, the laying of submarine cables, and the construction of docks and lighthouses. Other key topics include developments in industrial and manufacturing fields such as mining technology, the production of iron and steel, the use of steam power, and chemical processes such as photography and textile dyes.

The Life Story of the Late Sir Charles Tilston Bright, Civil Engineer

Sir Charles Tilston Bright (1832–88) was a renowned telegraph engineer, best known for his role in laying the first successful transatlantic cable in 1858, for which he was knighted. Bright later worked on the telegraph networks that would span not only the British Empire but the entire globe. Written by his brother Edward Brailsford Bright (1831–1913) and son Charles (1863–1937), both telegraph engineers who worked alongside him, this two-volume biography, first published in 1898, would do much to cement Bright's reputation as an electrical engineer, providing an insider account of telegraphy's formative years. Volume 1 traces Bright's career as an early employee of the world's first public telegraphy company, the Electric Telegraph Company, and his work on land and submarine cable telegraphy, culminating in the laying of the first transatlantic telegraph cables in the mid-nineteenth century.

Cambridge University Press has long been a pioneer in the reissuing of out-of-print titles from its own backlist, producing digital reprints of books that are still sought after by scholars and students but could not be reprinted economically using traditional technology. The Cambridge Library Collection extends this activity to a wider range of books which are still of importance to researchers and professionals, either for the source material they contain, or as landmarks in the history of their academic discipline.

Drawing from the world-renowned collections in the Cambridge University Library and other partner libraries, and guided by the advice of experts in each subject area, Cambridge University Press is using state-of-the-art scanning machines in its own Printing House to capture the content of each book selected for inclusion. The files are processed to give a consistently clear, crisp image, and the books finished to the high quality standard for which the Press is recognised around the world. The latest print-on-demand technology ensures that the books will remain available indefinitely, and that orders for single or multiple copies can quickly be supplied.

The Cambridge Library Collection brings back to life books of enduring scholarly value (including out-of-copyright works originally issued by other publishers) across a wide range of disciplines in the humanities and social sciences and in science and technology.

The Life Story of the Late
Sir Charles Tilston Bright
Civil Engineer

VOLUME 1

EDWARD BRAILSFORD BRIGHT
CHARLES BRIGHT

CAMBRIDGE UNIVERSITY PRESS

Cambridge, New York, Melbourne, Madrid, Cape Town,
Singapore, São Paolo, Delhi, Mexico City

Published in the United States of America by Cambridge University Press, New York

www.cambridge.org
Information on this title: www.cambridge.org/9781108052887

© in this compilation Cambridge University Press 2012

This edition first published 1898
This digitally printed version 2012

ISBN 978-1-108-05288-7 Paperback

The Life Story of
Sir Charles Tilston Bright

CIVIL ENGINEER

F JENKINS HELIOG PARIS

CHARLES BRIGHT

WHEN KNIGHTED   AGE 26

The Life Story

OF THE LATE

Sir Charles Tilston Bright

CIVIL ENGINEER

WITH WHICH IS INCORPORATED THE STORY
OF THE ATLANTIC CABLE, AND THE FIRST
TELEGRAPH TO INDIA AND THE
COLONIES

BY
HIS BROTHER
EDWARD BRAILSFORD BRIGHT
AUTHOR OF "THE ELECTRIC TELEGRAPH" "VIS" ETC

AND
HIS SON
CHARLES BRIGHT F.R.S.E
AUTHOR OF "SUBMARINE TELEGRAPHS" "SCIENCE AND
ENGINEERING DURING THE VICTORIAN ERA," ETC

VOLUME I

WESTMINSTER
ARCHIBALD CONSTABLE AND CO
2 WHITEHALL GARDENS

BUTLER & TANNER,
THE SELWOOD PRINTING WORKS.
FROME, AND LONDON.

TO

His Widow

THIS WORK IS DEDICATED

IN TOKEN OF AFFECTION

BY THE AUTHORS

Preface

IN the first place we wish it to be understood that in undertaking the biography of the late Sir Charles Bright ourselves, instead of entrusting it to a writer more accustomed to work of this nature, we do so in the belief that though our literary powers may not be worthy of our subject, we can nevertheless give a more accurate and true story of the man and his life and work : just as a writer who has actually witnessed a scene which he describes, pictures it often in a more vivid manner than another who, not an eye witness, may be in a literary sense his superior.

No one it seemed to us was better qualified in this case than his brother, who was his fellow worker and collaborator throughout the greater part of his life; in conjunction with the son, who has, for the last fifteen years, endeavoured to follow in his father's footsteps. In our opinion the most striking memoirs have been almost invariably published either as autobiographies, or else have been written by some one intimately associated with the subject of the memoir throughout his lifetime.

We have both of us been animated throughout by one motive only, namely, to put together, in a readable way, the great developments of science in which the subject of this biography took so active and prominent a part; and to describe—in what we hope may prove an interesting form—his achievements in the many, varied, and vast undertakings with which he was connected, and to which he devoted the greater part of his life.

In effecting this object, we have made reference to a quantity of early publications and original documents, official and otherwise, emanating from various quarters; and where this has been the case we acknowledge our indebtedness either in a footnote or in the body of the text.

We wish, however, to take this opportunity of expressing our especial indebtedness to *The Times* newspaper and *The Illustrated London News*. To Lord Kelvin, LL.D., F.R.S.; Sir C. L. Peel, K.C.B.; Sir E. J. Reed, K.C.B., F.R.S.; Maj.-Gen. Sir Frederick Goldsmid, K.C.S.I., C.B.; Captain Sir Allen Young, C.B.; Sir Henry Mance, C.I.E.; Sir W. H. Russell, LL.D.; Mr. J. C. Parkinson, J.P., D.L.; Mr. John Penn, M.P.; Sir Samuel Canning; Mr. W. H. Preece, C.B., F.R.S.; Mr. Latimer Clark, F.R.S.; Mr. F. C. Webb; Mr. Henry Clifford; Mr. Robert Kaye Gray; Captain A. W. Stiffe, R.I.M.; The Hydro-

PREFACE

grapher to the Admiralty, Rear-Admiral Sir W. J. L. Wharton, K.C.B., F.R.S.; Commander Douglas Gamble, R.N. (Naval Intelligence Department, Admiralty); Mr. Robert Dodwell; and Dr. W. S. Porter, amongst many others, for valuable information and interesting particulars, besides H.M. Post Office, The India Office, The Anglo-American Telegraph Company, The Eastern Telegraph Company, The Telegraph Construction Company, The "Silvertown Company," Henley's Telegraph Works Company, and other corporate bodies.

We also desire to acknowledge the cordial co-operation of the Publishers in the production of these volumes.

The order of date has been observed as far as possible with due regard to continuity of subject.

In several parts the work has had to be prepared without consideration of the fact that one of the authors was closely identified with many of the undertakings in question.

In reference to the scope of the book, the reader should bear in mind the full title as set forth on the title page. It was felt that, whilst practically telling the story of pioneer submarine telegraphy, it would have been a pity if the connecting links thereof were not as nearly complete as possible, compatible with the main object in view. Thus we have also dilated on matters with which the subject of our memoir was less intimately connected.

Introduction

WITH us it is naturally a work of love to chronicle the exploits, inventions, and scientific achievements of one who, when a youth of nineteen, carried out important telegraph work in Lancashire and Yorkshire; when only twenty became Chief Engineer to the Magnetic Telegraph Company, extending its lines throughout the United Kingdom ; and who was the first to connect Great Britain and Ireland by submarine cable, on which occasion he was accompanied on the expedition by his bride, to whom this memoir is dedicated.

When only twenty years of age he had patented, together with the brother constantly associated with him, no less than twenty-four distinct inventions, which had been elaborated during several preceding years, and many of which are still in use and essential to the telegraphy of the present day.

When but twenty-three he became one of the projectors, together with Mr. Cyrus Field and Mr. J. W. Brett * of the Atlantic Telegraph of 1856,

* Joined later by Dr. William Whitehouse.

xi

to which he was appointed Chief Engineer when only twenty-four years old, and which, after a series of almost insurmountable difficulties, was successfully laid by him in 1858 between Ireland and America, at the age of twenty-six. It had been said by many who had watched his energy and talent in early days, that honours were in store for him. The prediction was verified. He was knighted in that same year for what was at the time very justly described by the Press as "the great scientific achievement of the century."

His youthful talent has been spoken of as scarcely second to that of William Pitt.

After carrying out many important works in the Mediterranean and elsewhere, including the first telegraph to India, he was elected Member of Parliament for Greenwich at the age of thirty-three.

He continued his active career of practical work and invention in various departments of engineering—telegraphy, electric lighting, etc.—until his death in 1888. He filled the post of President of the Society of Telegraph Engineers and Electricians (now the Institution of Electrical Engineers) during the Jubilee year of 1887.

There are not many cases in which so much useful and important work has been done in so short a time and at so early a period of life. Indeed *The Times*, in its leading article on the occasion of his death, remarked :—" If a man's life may be measured by the amount he has accomplished, Sir Charles Bright lived long, though dying at the comparatively early age of fifty-five. Few men have ever done more useful work for his country and for commerce in the short space of twenty-seven years." A glance at *Men of the Time* and *Men of the Reign* will substantiate this statement.

There is probably no branch of engineering which lends itself so readily to a full sight of the world as that of Telegraphy. Apart from his profession, however — in his varied tastes and recreations—Sir Charles was perhaps as much the traveller as the scientist.

His life was throughout fraught with danger and anxiety. Essentially a man of action and endowed with great ability, his main characteristics were, we consider, intense energy and keenness, patience, fortitude under adverse circumstances, determination, perseverance and resource. He seemed constantly to be living up to Longfellow's lines :—

> " Each morning sees some task begun,
>   Each evening sees it close ;
> Something attempted, something done,
>   Has earned a night's repose."

Another conspicuous feature was his wide and varied scope of interests, sympathies, and friends.

It has often been said that a biography should open with a reference to progenitors, with the object of determining whether any predecessors have evinced special characteristics developed in turn by the subject of the biography. By thus feeling the "ancestral bumps" (phrenologically speaking) we often establish an "hereditary trait" not without interest.

Where the ancestry cannot be traced, the youthful beginnings may themselves be investigated, in order to see if early predilections for Robinson Crusoe, cricket, etc., appear to have led to the later development which has rendered the life a notable one. But schoolboy peculiarities seldom give much sign of the future, and mostly, —like Artemus Ward's frog—"don't show nothing particular."

In the families from which Sir Charles originated the traits we mostly find are combativeness, religion, and literature, mingled with legal and medical astuteness, and there is no accounting by antecedents in this case for any predisposition to science or invention.

Contents

CONTENTS

List of Illustrations

LIST OF ILLUSTRATIONS

xviii

LIST OF ILLUSTRATIONS

Folding Sheets

xix

Chapter I

FAMILY MEMOIRS

THE subject of this biography was descended from the ancient Hallamshire family of Bright, of which he represented the senior branch. The Hundred of Hallamshire comprises a considerable district of Yorkshire around and embracing Sheffield. In this division [1] the family were formerly owners of the Manors and Halls of Whirlow, Graystones, Carbrook, Badsworth, and Bannercross. In Sheffield itself there is a densely populated part named Brightside,[2] about which a tradition is current, that it was originally Bright's "hide"—*i.e.* a "hide" of land belonging to the family. Or, again, "the side of the town belonging to the Brights" is sometimes given as the origin of the town's name.

[1] Partly constituted by, and embracing, the manor of Ecclesall and the township of Ecclesall-Byerlow. In early days the village of Brightside was some three miles from Sheffield.

[2] For many years represented by an old friend of Sir Charles, *i.e.* the Right Hon. A. J. Mundella, M.P., lately deceased.

SIR CHARLES TILSTON BRIGHT

Probably no history of Sheffield or of Hallamshire, the "hundred" surrounding the town, has ever been written (or could be complete) without a considerable part being devoted to the Bright family who had estates all round the district—at one time perhaps the most beautiful part of Yorkshire—and who have occupied very distinguished positions in the town and neighbourhood.[1]

The most notable member of the family in ancient times was Colonel Sir John Bright, Bart., of Carbrook Hall and Badsworth,[2] in the time of the Commonwealth, who was a military chief under Cromwell, and fought with Lord Fairfax in the Parliamentary wars. He raised a regiment of horse on his own estates, and was in turn Governor both of York and Sheffield, afterwards succeeding Lord Hotham as Governor of Hull. He was also a High Sheriff of Yorkshire. During the Commonwealth Colonel Bright was one of the six representatives in Parliament of the West Riding, and an active member of various committees in the country. When, however, the execution of King Charles was decided upon, he

---

[1] See *Sheffield: Past and Present*, by the Rev. Alfred Gatty, D.D. (London: Bell & Sons).

[2] Now the property and home of Mr. R. H. Heywood-Jones, J.P., whose guest here one of the authors has had the great pleasure of being. From a photograph taken by the Rev. R. Y. Whytehead, Vicar of Campsall.

BADSWORTH HALL

Restored about the middle of last century, and formerly the seat of Colonel Sir John Bright, Bart., 1619–1688

3

withdrew from the Parliamentarian ranks, and disbanded his regiment. He subsequently assisted in the Restoration, and was created a baronet. On his death a handsome marble monument was erected to the memory of Sir John Bright in Badsworth Church, and, being of some historic interest, a view of it is given here. Though a party to four separate marriages—all into families of his own county—this distinguished Parliamentarian commander had only one son, who pre-deceased him but a few days, after a union with the Lady Lucy Montague, daughter of Edward, Earl of Manchester. The only surviving child of Sir John Bright married Sir Henry Liddell, Bart., of Ravensworth Castle, connected with the Duke of Grafton and Lord Ravensworth.[1]

---

[1] Hunter's *South Yorkshire* (p. 14) says : " The Puritan families of Westby, Bright, Spencer, Gill, Hatfield, and Staniforth, united at once in blood and in a community of political and religious feeling, possessed an interest and an influence at and around their country which nothing but military power could countervail."

In his *Hallamshire* also, Mr. Hunter observes : " The Brights of Carbrook and the Spencers of Attercliffe were decided Parliamentarians " ; and he adds, " the house of Howard had not an influence sufficient to counterbalance that of the two last named families, aided by what was the general feeling." For further particulars see the newly published *Dictionary of National Biography*, vol. vi. (Smith, Elder & Co.), wherein is a memoir of the late

The eldest son by this marriage, Thomas, was the ancestor of the first Lord Ravensworth. John, the second son, was made principal heir of his grandfather, on whose death he assumed the name and arms of Bright. At one time M.P. for Pontefract, he was the originator and first master of the Badsworth Hunt which, in connection with the said John Bright's mastership, boasts the oldest hunt song in existence. A copy of the original is reproduced in the appendices at the end of this volume, an old copy having been kindly lent to the authors by Mr. R. H. Heywood-Jones of Badsworth Hall. It may be observed that the song is styled "descriptive of an excellent Fox Chase, as performed by the hounds of Mr. Bright, of Badsworth, in the year 1730." History has it that Bright was wont to suffer from the gout—a malady which appears to have run in the family—and occasionally he might be seen following his hounds in a Bath chair! There is a coloured chart in the possession of Mr. G. B. Cooke-Yarborough, J.P., of Camps Mount, near Doncaster, in illustration of a run on the part of the Badsworth Hunt, which actually depicts Master Bright so esconced.

A grand-daughter of the above, Mary Bright,[1]

Sir John Bright. A full account of his funeral is given in Boothroyd's *History of Pontefract*.

[1] Whose cousin married Lord John Murray, of the 3rd Regiment of Foot Guards, a son of the Duke of

MONUMENT IN MEMORY OF SIR JOHN BRIGHT, BART.

was married to the Marquis of Rockingham, who was Prime Minister for a short time towards the end of last century. She survived her husband for a number of years, and the Carbrook and Badsworth properties, near Pontefract, worth many thousands, passed, for want of issue, to the Fitzwilliams. A portrait of the Marchioness, who was very beautiful, is given in Hailstone's *Yorkshire Worthies*.

Among Sir Charles' uncles was Major Henry Bright, of the 87th regiment, who was at the storming of Seringapatam when Tippoo Sahib was killed, and who himself fell later whilst

Athol. She was the grand-daughter of John Bright, of Bannercross, near Sheffield. For further information regarding this branch of the Bright family those Yorkshire antiquarians who are sufficiently interested should consult *The Bagshawes of Ford*, by W. H. Greaves-Bagshawe, J.P., D.L. (London : Mitchell & Hughes). The Bagshawes were cousins of the Brights of Carbrook and Badsworth. Moreover, in former days (1661), Grace, daughter of Henry Bright, of Whirlow Hall, in the parish of Ecclesall, married Mr. John Bagshawe, of Hucklow Hall, and High Sheriff of Derby. The family of Greaves were also connected by marriage with the Brights. Banner Cross, the property for a number of years past, of Mr. Greaves-Bagshawe aforesaid—and now tenanted by Mr. Douglas Vickers—is one of the finest surviving specimens of old Yorkshire estates. It was connected with the Bright family for many years — from the time of Elizabeth—and Sir Charles once visited there.

leading his regiment at Toulouse—the late Field Marshal, Lord Gough, being at the time his junior major. It may here be remarked that if the electric telegraph had then existed this last battle would probably not have been fought, for Napoleon Buonaparte having already abdicated at Paris, Wellington and Soult would have known that the articles of peace had been signed between England and France some days before.

Another uncle was Lieutenant Thomas Bright, R.N., who died from an accident on board of the *Fighting Arethusa*. A third was the late Dr. John Bright, of Manchester Square and Overton Hall, Ashover, an estate which he acquired from the representatives of Sir Joseph Banks, the famous naturalist of Captain Cook's expeditions. The said Dr. Bright acted as physician to George III. during a great part of his reign. A son, the late Rev. Mynors Bright, became principal of Magdalene College, Cambridge, and was known in the world of literature for his exhaustive and scholarly edition of *Pepys' Diary*, about which there was much controversy with Lord Braybrooke, who had previously brought out a much less complete edition.

Another uncle was the late William Bright, Town Clerk of Doncaster and agent to the estates of Earl Fitzwilliam. His son, the Rev. William Bright, D.D., Canon of Christ Church, and Regius Professor of Ecclesiastical History at

Oxford University, is distinguished as a writer of religious poetry and other works, including some of the most beautiful verses in our Ancient and Modern hymn book.

The pedigree of all branches of the Yorkshire Brights was given, with great ramifications, in the early part of the century, in Hunter's *Hallamshire*.[1] More recently a condensed form, with reference to this branch of the family, was published in Burke's *Authorised Arms*,[2] which is appended to the present volume (see Appendices).

The late Sir Bernard Burke in his description says : " The family of Bright is one of antiquity and distinction," and he—as well as Sylvanus Morgan, in *Armilogia*—expressly refers to the coat of arms as an excellent instance of "armes parlants," or arms indicating the name of the bearer by their actual characteristics. To readers conversant with heraldry, these armorial bearings may prove interesting. A facsimile is therefore reproduced.

---

[1] *The History of Hallamshire*, by Joseph Hunter, F.S.A., edited by the Rev. Alfred Gatty, D.D. (London : Virtue & Co., 1869). The first edition came out in 1819.

[2] *A Selection of Arms Authorised by the Laws of Heraldry*, by Sir Bernard Burke, C.B., LL.D., Ulster King of Arms (London : Hamilton & Sons, 1863). See also early editions of Burke's *Landed Gentry*.

Description

Shield. Parti per pale azure and gules, bend dexter or, between two mullets argent.

Crest. The sun issuant from a mass of clouds, all ppr.

Motto. "Clarior e tenebris."

The heraldic interpretation of the above is : azure (blue) division of the shield represents *morning*, and the gules (red) half *evening* ; the bend dexter or (golden band) is significant of the *sunlight* passing across the day ; while the mullets argent (silver stars) on each side are typical of the morning and evening stars. All the above are emblematical of the name ; confirmed by the crest of a sun issuing from behind clouds, when, as the motto explains, it shows " brighter out of darkness." [1]

[1] It is somewhat curious that the arms of a family of distinction of the same name in the West of England are

𝕽ame. Sixty years ago the surname " Bright," was very uncommon, though of pleasant significance, but since the advent into public life of England's great orator, John Bright, the " Tribune of the people," his fame has much multiplied the name, by inducing a number of those whom he would have termed " the lower middles," to discard their original patronymics for his, though perhaps mostly in the case of individuals whose better designation might rather be " Dull," or, at all events, " Unpolished."

Sir Charles was the possessor of a most curious carved oak mantelpiece—which came originally from his grandfather's old place, Graystones—perfectly black with age, and carved in 1672 with effigies of Thomas Bright,[1] of Graystones,[2] and his first and second wives, and upon each of the recessed panels is a representation of the family crest—a rising sun.

He also possessed several other trophies con-

identical, with the sole exception that the left (sinister) half of the shield is gules, and the right (dexter) half azure. Most probably the difference was made in ancient days to distinguish the families, but it also forms a singular coincidence with the narration of the diurnal reference to creation in Genesis.

[1] Second son of Henry Bright of Whirlow and Ecclesall.

[2] Graystones—or Greystones, as it is sometimes spelt—is now tenanted by Alderman Michael Hunter.

nected with his distinguished lineage, all of which are still held by the family.[1]

A recent *Nineteenth Century* article by Dr. Conan Doyle, on " The Distribution of British Intellect," classified the subject of our biography amongst the East Saxons of Essex. Surely, however, if any argument is to be derived from analyses of this description, the accident of birthplace should take quite a secondary position as compared with that of locality of ancestors, early youth, and education. Under the first heading—if for no other reason—Sir Charles should be classified as a Yorkshireman rather than anything else.

---

[1] In addition to Hunter's *Hallamshire*, Foster's *Pedigrees of the County Families of Yorkshire*, and Hailstone's *Yorkshire Worthies* aforesaid, the genealogical and antiquarian reader is referred, for further particulars on the subject of this chapter, to *Notes on the Hallamshire family of Bright* by William S. Porter, M.D., which appeared originally in the first number of the *Sheffield Miscellany* (December, 1896, and January, 1897), forming the first of a series of articles on Yorkshire families. Dr. Porter is related to the Bright family,—as well as to the Taylors,—and has always shown a conspicuous interest in matters concerning antiquarian Yorkshire.

THE GRAYSTONES OVERMANTEL

15

Chapter II

BOYHOOD

BORN on June 8th, 1832, Charles Bright, the youngest son of Brailsford Bright, was brought up with his brother Edward, and this period of his life is perhaps best described in the brother's own words :

Our early life was not only happy but healthy. We lived on the verge of Hainault Forest, with a large garden about the house over which we could ramble to our heart's content, and, except when at lessons with our tutor, we were all day in the open air.

Our father was an ardent sportsman of the old sort, with a strong taste for riding, fishing and shooting. In his earlier days he was termed by many of his acquaintances " Beau Bright," because of his extreme neatness of attire and love of society. An enthusiastic dancer, he frequented the Caledonian and other societies associated with Willis's Rooms. He was also a most genial host. Later on, becoming somewhat corpulent, and

17   C

habitually riding a very big roan horse into town, his nag and he were dubbed the "Elephant and Castle."

In later days he lived at the seaside, at Rhyl, both for the sake of his health and that we might easily visit him from Liverpool. He died there in his eighty-fifth year, our mother having predeceased him by fifteen months.

Our mother was the only child of Edward Tilston, of Mold. Tilston was also possessed of property at Wellwyn and Upton, where he died. He had been on friendly terms with Lord Nelson, who became a sponsor to his daughter, by whom she was named "Emma Charlotte." Nelson gave a beautiful ring for her at the time, with his coat of arms, inscribed "Nelson and Bronté."

She was tall and graceful, an accomplished musician, playing equally well on harp and piano; moreover, she had a very sweet voice.

Tilston was another old Yorkshire family, whence the subject of this memoir derived his second name.

We were constantly on pony back when children, and fished as soon as we could hold a rod. There was a large irregularly-shaped pond in a meadow adjoining the property, partly surrounded with trees; and in addition to many weeds and bulrushes we found, as young fishermen will, that it held plenty of roach and excellent carp. Indeed, so secluded was this nook (though not far from the house), that our father occasionally shot snipe from it, and in one winter bagged a woodcock. The trees being on the south side from which we fished, prevented the shadows falling on the water, and many a good creel full did we win by the seductions of our father's recipe of paste and honey with a little cotton wool added.

These carp were not like some too well fed, lazy, dyspeptic things, but long and muscular fish in full piscine vigour, and ready to fight hard. And now came the first instance I can call to my mind illustrating my brother Charles' power of invention and resource in an emergency. Once and again the tackle was broken by an enormous fellow that we occasionally got a sight of—very dark in colour, and a very " dark horse " to us. His size was naturally magnified in the imagination of boys who were fishermen also, and we fancied him much bigger than he was likely to be in reality; but we had read of giant carp, and vowed this particular specimen must be at least fifteen or twenty pounds, always imagining that it was the same fish. One or other would get hold of him in a hole seven or eight feet deep, when, after a rush or two, he would start right away for a patch of weeds in the deepest water and evade us.

This objectionable and adverse treatment, though at longish intervals, probably had the effect of producing an early development of bumps of determination in addition to the inherited bumps piscatorial. At any rate we set our minds on tackling the monster by investing our pocket-money in two specially strong gut traces and hooks, destined for our fish friend, or fiend—for we had almost got to look upon him as such.

Taking as our motto " Carpe diem," we fished away sedulously for a long time, only catching, if anything, rather smaller, and certainly fewer, fish than usual. Early one morning, however, when we were both "trying the deeps," my light float, after swirling about a little, went quietly down and travelled towards the still deeper water. I struck, and then the usual rod-bending commotion ensued, but relying on the tackle I managed to turn him from his mad rush towards his weedy stronghold on the other side, having a desperate bit of play to and fro for some time. The fish got tired— and so did I. We had no landing net or gaff, and on sighting us, he again went off at score —this time into thick weeds to the left where it was much shallower—and then he would not move. We were now more than ever determined to have grandfather carp—but how? Haymakers were in an adjoining field. But it was decided not to ask their assistance, mostly for the sake of the honour and glory of effecting the capture off our own bat—or rather, rod—but partly also because we thought perhaps they might collar the big fish for themselves. Suddenly Charles cried, " Wait a minute, and I'll have him." Scrambling over a hedge and ditch, he borrowed a big hay rake, with which he waded into the water a little way off where it was fairly shallow. He had the greatest difficulty in getting through the weeds to the edge of the clump (where our foe was secluded), and was then up to his waist in water.

Just then our wily friend was probably startled, for he shook the weeds, showing his temporary habitat. Down Charles popped the rake on the other side. As it happened, Mr. Carp had so jammed himself into the mat of stuff in his desperation that he couldn't back out. Thus, the rake was got under him, and he was pulled up—hook, tackle and all. My brother then pressed a heap of weeds on the broadside of the row of big prongs, and so enclosed—and finally, landed—our friend !

We never had such a joy in fishing before or since, and, as may be imagined, were off home at once to announce our capture, which was found afterwards to weigh nearly eight pounds. There was then the question of stuffing it. Our father, an old flyfisher, didn't care for "coarse fish," but had it stewed for us, and ordered some port wine, as well as spice, to make it edible ; and so it was "stuffed" in a different way. I doubt if we ever again appreciated fish so much, and as for dreams, we lads were in a sort of sporting heaven. My brother Charles, by transforming a hay rake into a landing net in a few moments, did what few children at that age would have thought of.

In addition to the pond, the rivers Lea and Roden were often tried. We also had permission to fish in the preserved waters of Wanstead Park —then owned by the Hon. Tilney-Long-Pole-Wellesley—but went seldom, as it necessarily

A NOVEL "GAFF"

entailed "tipping" the keeper, whereas we were able to have our sport "gratis" elsewhere.

Our father and an old friend had shooting manors at High Beach and Warleigh. Here one or other of us used to be taken alternately in the four-wheeled dogcart, and allowed to carry a game bag. Spanking along behind a pair of thoroughbred horses on an early September morning, with the dewdrops sparkling in the sun on the grass and leaves, and one of the dogs joyously bounding along by the roadside, was always most delightful. After the drive came the excitement of seeing the birds or other game brought down. This, together with the pride of taking part in carrying the spoil, formed a great item of early happiness.

The shooting over well-trained dogs—like Father's " Juno " and " Spot," pointers, and " Romp," a setter —was far more interesting than the usual "battue" work of the present day. Accompanying these two steady old sportsmen —for they never shot with others — the dogs coming steadily to the point and backing one another up, and the quiet "down charge" for reloading, almost constituted an education in itself. In those days the muzzle loader alone was in use. Thus, before starting, our father would always deliberately — indeed almost solemnly—pat successively the four special pockets in his old fashioned shooting coat, pronouncing as he did so the shooting shibboleth of those days—Powder! Shot! Percussion-caps! Wadding !

At this early period of existence, we boys fancied we could shoot, having knocked over various small birds (whilst sitting) by careful aim along the barrel of a single gun, but later on we were successively undeceived. On returning one afternoon with my father, he allowed me to try to knock over a partridge. I aimed, as usual, along the barrel, and to my intense disappointment made several bad misses. My father merely smiled, and said, "That won't do, now just look hard at the next bird, and try to see its eye ; then pull the trigger steadily, and the gun will come up to the bird of itself." I did so, and was astonished to see my first partridge fall. After such schooling we soon became fair shots, having much of practice at times in our holidays, both at home and when staying with our school friend Alfred Turner, at his father's place, Chinting House, near Seaford, in Sussex, where furzes and rabbits abounded in the neighbourhood of the downs on his extensive holding. Lanes existed through the furze, and Mr. Singer Turner's pack of beautiful little beagles were often employed to drive the rabbits across from one patch to the other. At one long Midsummer holiday visit, some extensive galleries were being driven by the Royal Engineers, to blow down part of the high chalk cliffs east of Seaford, with a view to checking the rapid washing away of the shingle. Lieutenant Ward, R.E.,[1] was in charge. I well remember one day, when at a rabbit *battue*, he had about fifty shots, but could not bag one. He accounted for this in some remarks after dinner, by saying that like Nimrod "he was a mighty hunter," but his prey was man.

[1] Afterwards Major-General Sir Edward Ward, K.C.M.G.

BOYHOOD

We occasionally passed our holidays with another school "chum," John Stallibrass, at his father's place, Thorpe Hall, Southchurch, and there we were first "entered" into the mysteries of fox-hunting, at the age of twelve and thirteen respectively. It involved rather a long ride, for the meet of the Essex hounds was some way on the other side of Rochford. I was mounted, by our host the "Squire," on a young, half-broken colt, and after the hounds had "gone away," the colt went away with me, and getting the bit between his teeth, I could no more guide him than fly. This, indeed, I quickly did, for after negotiating with more or less uneasiness a variety of jumps, he swerved. I was caught in the breast by an overhanging branch, taken clean out of the saddle in the manner of Absalom, and came down a tremendous crack on my back in the bottom of a broad dry ditch on the other side of the hedge.

I remember seeing all the stars of the firmament; but luck was with me, for my nag was caught, and I was able, with Charles, to be in at the death, whereupon we were honoured with a couple of pads, which were long cherished by us as trophies.

In after years many a day's pleasant shooting— as well as fox-hunting—have we had with our same old school friend, now alas the late, John Stallibrass, over his estate, Eastwoodbury, and on Foulness Island. To get to Foulness, it was necessary to drive over to Shoeburyness, where a cart was waiting to take us over the sands, which had to be crossed at low tide on a certain track of shallow water (marked out by poles and bushes), the neighbourhood here being bordered by quicksands. The island, reached after about two and a half miles drive, was some miles long, and surrounded by a high sea wall grown over with a thick scrub. As a rule, the coveys and hares, when startled, betook themselves to this cover on the sloping banks, and were easily marked down. Thus, with a gun at the top of the bank, and the others beating through the small bushes, a capital day's sport was to be had. There were some excellent oyster beds in the inlets about the island, and, as Lord of the Manor, toll was always laid upon them by our host's farm bailiff, resulting in a hamper of the best, just untucked out of their beds, for our lunch at one of the quaint old brown-tiled farmsteads.

There was an occasional drawback to the freedom and enjoyment of holidays when at home, for our father not only loved sport, but was devoted to whist. For very many years his evenings were mostly given up to it, and among his friends he was often dubbed " King Whist." He occasionally stood in need of a fourth, and didn't at all appreciate dummy; thus every now and then our noses were put to the whist grindstone (when passing eleven and twelve), so as to fit us for the honourable position of a fourth. Instead of being out and about, we were, for the time, impressed into a study of old "Hoyle" and "Major A." on short whist (the then authorities), with the cards spread out before us. And worse was behind, for, when somewhat *indoctrinated*, one or other had to sit down to play in the evening till perhaps nearly midnight; and when "hauled over the coals" for some blunder with parental admonitions, we heartily wished every card somewhere else.

When with our father in after days, we always joined at a game of whist in the evening. On one occasion, when he was over eighty, having just dealt and being called away from the table, we took advantage of his absence, and put all the trumps in his hand. On taking up the cards and sorting them, he looked narrowly at us, but being apparently unsuspicious, he struck his hand on the table, exclaiming, "You really need not play, my dears! For more than fifty years have I held cards, but *never* such a hand before!" We thought so too. He often referred to it, "speering" at us inquisitively; but we always held our faces (and our peace), to his comfort be it said.

If I have been led to dilate on these subjects, I can only say that field sports, and even cards, were of much more real utility to us in later days than all the Latin and Greek with which we were crammed for years. Even this knowledge of whist and other games has often helped materially to while away tedious moments on board ship when actual work was not going on, and when tired both of reading and deck-prowling.

I hope, also, I may be excused for dwelling somewhat extensively upon the early development of our sporting knowledge. It really proved of the greatest advantage to both of us in various ways in wild and deserted parts of the world. Thus, it often procured a salutary change of diet from ordinary ship's provisions, and sometimes, indeed, kept the wolf from the door when other provisions had failed.

School Days.—After a quiet country boyhood and tutordom, including many piano lessons, we were called upon, in the ordinary course of things educational, to make our plunge into a public school. Merchant Taylors' was chosen for us, there being family connections on the governing body. With regard to school days, our principal reminiscences have always been that it was then the "whacking age"! The birch had been discarded for the cane, and this was liberally applied to the hands for a false quantity in Latin, or a wrong number or tense. Frequent whacking naturally engendered temper, and, I believe, developed the bump of pugnacity. At all events, few days passed without a "ring" in the cloisters, to settle small questions of comparative supremacy by fisticuffs and more bumps. In school we became moderate scholars.

Latin and Greek, with a bit of Hebrew, we studied, whilst mathematics in large doses were indelibly impressed upon us. Outside this we obtained a fair education in self-defence, which, in after life, proved over and over again most useful to Charles and myself. As at most of the larger public schools, a great deal of freedom was allowed on afternoons. At first we concentrated most of our attention on rackets, and afterwards on rowing ; but our father stopped water peregrinations altogether till we had secured certificates from a master, to whom we were sent, for "proficiency in swimming and diving." A few weeks did it.

Shortly afterwards it was forced into our young heads, where it remained ever after for the best of reasons, that the parents of every child ought to be compelled by Act of Parliament to make them learn natation, and that there should be national baths and instructors for the purpose. I question whether any law would save more valuable lives at less cost; for those going afloat, whether at sea or on our rivers, are among the pick of our British flock.

In those days Merchant Taylors (one of the oldest public schools) held a somewhat similar position in boating matters to Eton, and had the benefit of the best of Oxford coaches. Thus, starting at an early age with a "free mind," we soon learnt to row in fair form, till in the spring of 1844 we were able to pull together in one of the first outrigged pair oars launched by Noulton and Wyld (then Searle's competitors) from their boat yard just above old Westminster Bridge.

At that time, owing to our fishing, swimming, and rowing proclivities, we were both fast becoming "water rats," or "wet bobs." I was a trifle older, and a bit the tougher of the two, though we were much of a size, and often boxed together.

The taste of Thames water was learnt before long, though not appreciated, for we were swamped before a month of our outrigger experiences by the swell of a steamer, and, in all, were obliged to swim for our lives about eight times. A boat collision accounted for one, Battersea Bridge for another. The old bridge there had very small openings between wooden piles, scarcely leaving room for oars, and in a pair outrigger (necessarily without a coxswain) it was very easy to come to grief which we did in "shooting" it once, when Charley, who steered as bow oar, missed his line, and my oar coming in contact with the bridge, over we went.

On another occasion we had pulled second in the Putney Regatta pair oars, and were turning into Avis' Inn (the rendezvous by Old Putney Bridge) when a steamer, accompanying the race, had too much way on, and ran into us. Her cutwater pushed the oar against my chest and sent me right over, also capsizing the outrigger. I sank like a shot, and just saw the red gleam of the stopped paddle float over my head before I came up astern and swam to a boat. Charles got hold of the forward chain, and was hauled up on the sponson. As he was shaking the water off, closely resembling a drowned rat, an old gentleman enquired, "May I ask, young gentleman, if you're insured." Charles always afterwards vowed that, from the old boy's prompt query, he must have been a director of a Life-Assurance Company.

But the most remarkable upset we ever experienced was in the school eight. We had rowed to Putney, one of our party, Dan Gossett, pulling up in a wager boat. He damaged an outrigger, and, as in those early days, the bow and stern had covers of thin mahogany that were detachable, we took off the one in front to make room for Gossett as a passenger home. This necessarily depressed the bow of the eight somewhat materially, while leaving it unprotected. It was a dangerous thing to do, though not occurring to us in that light at the time. However, as we passed Pimlico Pier, rowing with the tide in the middle of the river, we rushed into the swell of a passing steamer which at once swamped us. Charles and I collared our pea-jackets, which held our available cash, and struck out for shore. One of our crew, Church, couldn't swim; but the boat, relieved of our weight, floated upside down, and he sat himself on it and shouted "Rule Britannia"! Russell, a brother of one of our junior masters (afterwards head-master), could only paddle, so we brothers gave him a helping hand, but had in consequence to resign our peajackets to the mercies of the deep. Some were hauled on to a barge, some got ashore; our "eight" was towed to the bank, and the oars recovered by another boat. We then turned her over, cleared the water out, and rowed back to Noulton's. Here we changed, and, after a little dose of "hot and strong," were none the worse for our "ducking" in Thames water, though at that time, as now, of very objectionable consistency and constitution.

The marvel to my brother and myself ever afterwards was how ten lads of fourteen to seventeen could all get safe out of an almost instantaneous swamp in the middle of the Thames, off Pimlico, on a quick ebb tide. The sinking of the bow of our long eight was indeed so sudden that Charley Foster, our coxswain, had actually shouted out, "What are you doing Bow and Two," when the water was already on them!

Of this crew, in after days one became a captain of a famous ironclad battleship in our Navy, and has seen much service. His brother, a rector in Devonshire, was No. 3; No. 4 became a bank manager; another a vicar in Middlesex; No. 5, is in the Foreign Office; and our coxswain in the Indian Civil Service. My brother and I are accounted for by these pages, and the sitter in the bow, who sank us by his undue weight, is now mayor of a town not very far from this mighty brickfield called London, and chief of the local magistrates, after being for some time member of the Legislative Council in Queensland.

In those days the racing course was between Vauxhall Bridge and Putney Bridge, though the annual Thames Regatta was held higher up, between Putney and Chiswick. The four of the school was manned by some of the monitors and promptors (the seventeen to eighteen year olders), who were next on the "roster" for scholarships and exhibitions at St. John's, Oxford.[1] At that time we were comparative juniors in the fifth and sixth forms. The average age in our (second) four was about fifteen years; but we had continually practised, and rowed in a much lighter outrigger, so we challenged the first four, the course being from Vauxhall Bridge to Putney. They had greater weight and strength, and led at first. However, we were even with them

---

[1] St. John's College, Oxford, used in those days to be almost entirely composed of scholars from Merchant Taylors' School, Sir Thomas White, who founded Merchant Taylors' in 1540, having also founded St. John's, Oxford.

shortly after passing Battersea Bridge, and then drew ahead, reaching the winning post at Putney Bridge as they were rounding the turn below Hurlingham.

For further details of school life at Merchant Taylors', are they not given—including all the intricacies and mysteries of a hunt through Jupiter's Hole—by the late Albert Smith?

At the end of 1847, when Charles was sixteen and I seventeen, we left, taking various classical and mathematical prizes.[1]   On the whole, Charles preferred classics and I mathematics; but, in the course of our studies, both of us became very fond of chemistry and mechanics.

---

[1] Both of us made it through life a practice to attend our annual school dinners whenever able, as well as those of the Merchant Taylors' Club.

Chapter III

LAND TELEGRAPHS

CHARLES BRIGHT and his brother Edward were intended for an Oxford career, but owing to heavy pecuniary losses on the part of their father, the serious and more immediately practical side of life had to be at once entered upon. As schoolboys, they had very much interested themselves both in electricity and chemistry. Thus it came about that, soon after its formation in 1847, they joined, when respectively fifteen and sixteen years old, the Electric Telegraph Company,[1] under the auspices of Mr. (afterwards Sir William)

---

[1] This Company was started by Mr. Cooke and Mr. John Lewis Ricardo, M.P., and began business in 1848 between London, Birmingham, Manchester, Liverpool, and other large towns. Considering that this was the first Telegraph Company formed, it is curious that so long a time had elapsed since the famous Cooke & Wheatstone patent of 1837. This is explained, no doubt in the fact that a reliable insulator was wanted, and this was not forthcoming until the application of gutta-percha to this purpose.

Fothergill Cooke,[1] who was a connection by marriage. Within a year of entering upon this new field, both boys became inventors, and much of their spare time was devoted to thinking over, discussing together, and devising practical improvements on the Cooke and Wheatstone system which they then had to work upon,[2] as well as to other inventions of an entirely different character.

In those days, patent fees of £150 had to be paid, in addition to the heavy charges to patent agents for drafting, drawing, and completing patents. So much as this the brothers could not afford, so they contented themselves, for the time being, with starting a joint invention book, kept under lock and key, into which they, from time to time, entered up drawings, descriptions, and dates.

[1] On several occasions afterwards, Sir W. F. Cooke took the opportunity of expressing the pride he felt in having been the means of introducing Sir Charles to telegraph engineering and electrical science.

[2] It may here be mentioned that the first telegraph instruments worked were merely vertical compass needles on a spindle, which were moved to the right or left by an electric current when conveyed in one direction or the other, through coils of insulated wire surrounding them, the variations of deflections forming different letters. From the simplicity of this arrangement, coupled with a natural freedom from getting out of order, it went by the soubriquet "*toujours prêt*"; but it was slow, and the waverings of the beats of the needle rendered it subject to frequent error on the part of the operators.

LAND TELEGRAPHS

These were afterwards, with several additions, embodied in the celebrated and oft quoted patent of October 21st, 1852,[1] particulars of which are

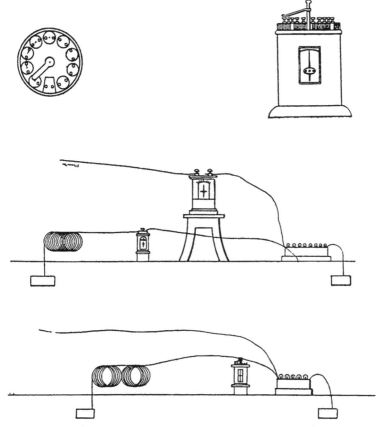

BRIGHT'S TESTING AND FAULT LOCALISATION SYSTEM, 1852

given in the Appendices to Vol. II., under "Inventions." It suffices to say here, that many of the

[1] Specification, No. 14,331 of 1852 ; E. B. and C. T. Bright.

novelties included therein are now in common use after a lapse of forty-five years.[1] Perhaps the most important of their inventions was their system—devised in February, 1849—of testing insulated conductors to localise faults from a distant point, by means of a series of standard resistance coils of different values, brought into circuit successively by turning a connecting handle. This drawing, reproduced from the 1852 specification (see Appendix), shows, even now, the best form of resistance coil arrangement in use for testing land and submarine telegraphs.[2]

It may fairly be said that capital would never have been found for the vast system of submarine cables throughout the world without the aid of this invention by the Brights, which enabled repairing vessels to at once go to the scene of damage,

---

[1] Youthful inventors may not be so very uncommon; but how many actually invent anything at the age of 17 which ever comes into practical use?

[2] This invention, in its various aspects, has sometimes been erroneously attributed to, or claimed for, others. Investigations instituted by Prof Fleeming Jenkin, F.R.S., in 1865, clearly established Sir Charles Bright's prior claim.

Mr. Jenkin found that the above patent constituted the first published description such as could enable others to test or localise, faults, by means of graduated resistance coils in a balancing, or differential, artificial circuit.

Besides the specification, this method was fully described in Lardner's *Museum of Science and Art*, of 1854.

instead of having to pick up and cut the cable at hazard at numberless points.

There was, however, one great advantage about a patent in those days, namely, that a number of separate inventions relating to the same general subject, such as " Electric Telegraphs," could be introduced. Nowadays, one improvement only is allowed, though (thanks to the Rt. Hon. Joseph Chamberlain, M.P.), at a cost of only £1 for the first nine months, and of £3 on the final specification, with a term of four years free from further charge, followed by small graduated annual charges.

The 1852 patent was the outcome of four years' brainwork of the brothers, while three years later there followed the patent of 1855.[1]

The year 1851 saw some important changes on the part of both the brothers. After having for some time been in charge of the Birmingham station, Charles left the " Electric " Company, and shortly after became Assistant Engineer to the British Telegraph Company,[2] which had lately been organised to work the patents of the brothers Highton, whilst Edward joined the Magnetic Telegraph Company, which had that year secured

---

[1] See Appendices.

[2] At first the British Electric Telegraph Company. Several companies had rapidly sprung into existence in response to a clamour for competition incited by the high tariffs of the original " Electric."

an Act for the exclusive right to lay overhead telegraph lines along the public highways.

Thus the two brothers became engaged in ad-
vancing the early stages of two competing concerns
—a curious and novel position for them. Charles's
headquarters were at Manchester, whilst Edward
was stationed at Liverpool. As a rule, however,
each passed alternate Sundays with the other.

Soon after joining the "British" Company, young
Charles addressed the following communication to
all the district superintendents and agents :—

British Electric Telegraph Company,
Central Offices, Royal Exchange, London.
Manchester Station, Market Street Buildings,
July 29th, 1851.

SIR,—With a view to assist you in replying to enquiries
that may be made respecting this Company, and for the
satisfaction of gentlemen in your district who may desire
to become shareholders, I beg to send you a condensed
account of its origin and progress to this time, together
with a carefully prepared estimate showing the data upon
which the directors ground their conviction that the undertaking will prove very handsomely remunerative to its
proprietors.

The British Telegraph Company originated as follows :—

In 1847 Mr. Edward Highton—at that time the resident
engineer of the Taff Vale Railway—was invited by the
Directors of the London and North Western Railway Co.
to relinquish that appointment in order to take the superintendence of their Telegraphic arrangements, he being
well-known as a first-rate practical Electrician—having, in

connection with his brother, the Rev. Henry Highton, the second master of Rugby School, devoted many years to the study of that science. The Board of the London and North Western Company were at that time exceedingly annoyed, both at the behaviour of the old Telegraph Company, and at the inefficient working of their instruments, and in consequence requested Mr. Highton to devote his entire energies to the discovery of such instruments and principles in Telegraphs as might be used without infringing the patents of the Existing Company. He at once took up the subject, and after lengthened investigations and experiments completely succeeded. The instruments and batteries thus invented were exhibited to the Society of Arts and obtained the large gold medal of that Society. The Directors of the London and North Western Company thereupon ordered them to be laid down over the whole of their lines. At the earnest request of and in consequence and consideration of a new and improved arrangement with the old Telegraph Company, this order was never fully carried out, but the Directors of the London and North Western insisted upon having the telegraphs of Messrs. Highton at all the most dangerous tunnels and junctions, and upon the Northampton and Peterborough Branch—then a single line—at these places, in all thirty stations. These Telegraphs still remain, having worked most satisfactorily during three years, and being at the present time under the superintendence of Mr. Highton.

In 1849 several gentlemen known to Mr. Highton, having become aware of the value of his inventions and being convinced from a careful investigation of facts that the business of a Commercial Telegraph was extremely profitable, entered into an agreement with him to join in the expense of obtaining an Act of Parliament. In 1850, in spite of severe opposition by the old Telegraph Company, they succeeded, and in 1851—after the perfect formation of the Company, and after having (on January 18th) obtained the written permission of the London and North Western Company to construct Commercial Telegraphs over the whole of their lines—the prospectus of this Company was issued and the public for the first time asked to join the undertaking. The Directors of the London and North Western Company, tempted by immense pecuniary offers made to them by the old Telegraph Company for the exclusive use of the line, have since refused to carry out their written permission as above, though the very terms they dictated were accepted by the British Company. A long time having been spent in the vain endeavour to induce them to act up to their honourable engagement, negociations have recently been opened with other Companies, and an arrangement on favourable terms for the British Company has lately been concluded with the Lancashire and Yorkshire Company by which we obtain Railway access of the best kind to the following important towns in those districts — viz. Liverpool, Manchester, Leeds, Bradford, Halifax, Huddersfield, Oldham, Ashton, Rochdale, Bolton, Wigan, Preston, Fleetwood, Blackpool. Southport, Bury, Blackburn, Goole, Dewsbury, and Wakefield. The lines of Telegraph in this district will be immediately commenced between Leeds and Liverpool, and we are in daily expectation of concluding a highly satisfactory arrangement with the Great Northern Company by which the chain of communication will be forthwith complete to London.

The directors are also making arrangements with a foreign Telegraphic agency of high respectability with a view to establishing in the central offices of the British Telegraph Company a "foreign department," in immediate connection with all the chief cities and ports of the Continent, in each of which the parties have already agents for the collection and transmission of news. The result of this arrangement will be that messages can then be dispatched and answered in an incredibly short space of time, *viâ* Calais, to and between all the chief markets, ports, stock exchanges, and manufacturing districts in this country and the Continent. As this arrangement will be entirely on mutual principles between this Company and the agents, it will doubtless be a source of very great emolument to both parties.

The Company's Capital consists of four thousand shares of £25 each (or £100,000), with powers to borrow to the extent of one third of any amount that may at any time be called up for the time being. A Deposit of £2 10s. per share is to be paid on allotment, calls are not to be made more frequently than once in two months, and no larger amount than £2 10s. per share is to be called for at one time. Of this capital, nearly one half is either actually issued or arranged to be taken up, and our great point now is to obtain *interest* by securing for the undertaking the countenance and patronage of locally influential gentlemen as shareholders in all the great seats of our future business. I give you below for your information the names of a few of our largest shareholders in the northern part of England, to either of whom any gentleman with whom you may converse on the subject is at liberty to refer :—

Richard Carter, Esq., Blackboy Colliery, Darlington.
Joseph Craven, Esq. (Craven & Harrop), Bradford.
R. W. Jackson, Esq., Greatham Hall, Stockton-on-Tees.
Thomas Wood, Esq., Trindon House, Ferry Hill, Durham.
William Dawson, Esq., Hartlepool.
J. S. Challoner, Esq., Newcastle-on-Tyne.

SIR CHARLES TILSTON BRIGHT

As regards the expected income of the Company and the data upon which those expectations are founded, I beg to draw your attention to the enclosed statement of account. The moderation of which—from what you yourself have seen of Telegraph business—will, I am sure, at once strike you. The Capital and the Annual Expenses are considerably over-estimated, and the returns are taken at about *one third* of those actually experienced by the existing Company two years ago, at their very high prices. The well-known fact that the messages will increase when the price is lower, has not in this estimate been taken into account at all, nor have I allowed anything for messages sent between Leeds and Liverpool, and Leeds and Manchester, and *vice versâ*, nor have I reckoned as of any value intermediate stations, such as Halifax, Huddersfield and Bolton, which might be established in direct connection with the others ; and messages between Bradford and Liverpool, and Bradford and Manchester, and *vice versâ*, have not been reckoned—in spite of which the figures show a nett profit of nearly 20 per cent. per annum.

Trusting that these facts and considerations, when laid before some of your chief merchants, etc., may lead them to see the soundness and value of the undertaking,

I am, Sir,
Yours obediently,
CHARLES T. BRIGHT.

On taking up his new position, Charles Bright was at once engaged in superintending the erection of telegraphs for the British Telegraph Company on the Lancashire and Yorkshire and other railways, as well as in connecting and fitting up various telegraph offices.

LAND TELEGRAPHS

The following is a copy of a letter he wrote to the young lady to whom he was then engaged, and who shortly after became his wife :—

British Electric Telegraph Company,
London,
September 5th, 1851.

I received your letter yesterday but could not answer it, as I was fully engaged until past post time.

You may easily imagine that with 160 miles of line which I have to commence at once, and a great many more directly after—if not nearly at the same time—that I have a great deal to look after. The only person who could assist me, one of the directors, is fully engaged with bringing out a Bill for next Parliament for a new railway line. So I am the only manager of telegraphic detail for the campaign, in addition to which I have some twenty-five long patents to bear in mind as to their separate claims and intentions so as not to infringe any other people's property.

I look forward to a stormy and active life for the next six months in various parts of the country—a life which I shall go into with pleasure, as I have *you* as the prize to look and hope for. It will not be unpleasant to me—however uncomfortable generally and disagreeable in detail—for as you know, my aim for some time has been to weave a web of wire in opposition to the monopoly, and, as I cannot do it for ourselves, I am well content to do it for others. Having no stake or responsibility in it I feel more comfortable perhaps than I should have had we succeeded in establishing a Company, which would have been a case of either make or mar.

I write you these business details, dearest, because I know they will not be tedious to *you*, and because I think you may have wasted your thoughts in speculations as to what

I could be doing in London! . . . It is pleasant to be engaged in a work of interest to oneself, and how much more when there is an *object* to be worked for so dear as my own B——!

You will be glad to hear that there is nothing irksome or unpleasant in my position with the Company. Though very young (I haven't told them *how* young!) I am looked up to, and I have no reason to be dissatisfied at all. I am treated kindly and like a gentleman, and it is astonishing how much more energetically one can work with such treatment than with that *distance* which is so common between directors and officers of a Company. The promises held out to me at first have been renewed, and I hope I shall hold even a higher position than I was sanguine enough to anticipate; but of course I do not expect everything at once, or until the directors receive some return—or without some actual work and thought. . . .

CHARLES TILSTON BRIGHT
Age 19

Young Bright was first brought into public notice by a remarkable feat, namely, the laying of the Manchester underground wires in 1851.

LAND TELEGRAPHS

It was essential that the traffic of so busy a city should be interrupted as little as possible. Charles Bright did not interrupt the traffic at all. In one night he had the streets up, laid the wires, and had laid the pavements down again before the inhabitants were out of their beds in the morning. He was then but nineteen, and received great credit in the public journals, notably in *The Times*, which made this piece of work the subject of a leading article.

The following arrangements for the night's work go to show the prescience and energy characteristic of him. Iron pipes were cast in halves, longitudinally, with side tongues and clips. The gutta-percha covered wires were wrapped with tarred yarn into a rope, and wound on broad drums with huge flanges. A large number of navvies were engaged, with competent foremen. To each gang was assigned a given length of street, along which the flagstones were to be lifted, the trench opened to the requisite depth, and the underhalves of the pipes laid and linked at the bottom. Another gang at once followed, wheeling the drum (whose breadth exceeded that of the trench), and unwinding the rope of wires into the under-halves of the pipes previously laid down. A further gang followed for applying, linking, and tightening the upper-halves of the pipes, while yet another set of men filled up the trench and replaced the flags. This operation, though easily described, required at this early stage of telegraphy a great deal of consideration, coupled with very active and determined control throughout the short night. Charles Bright subsequently carried out the same system in London, Liverpool, and other large towns.[1]

To turn again to the "Magnetic" Company. On its success being demonstrated, mainly under the auspices of the late Sir Charles Fox and Mr. Robert Crosbie, capital was quickly forthcoming for the organisation of a powerful Chartered Company under limited liability, entitled the English and Irish Magnetic Telegraph Company. To the working up of this Edward Bright devoted himself. In addition to Sir Charles Fox and Mr. Crosbie, the original holders of shares included Sir Joseph Ewart, M.P. for Liverpool, William Langton, T. D. Hornby, Edward Cropper, Henry Harrison, T. B. Horsfall, Sir Thomas Brocklebank, Sir Hardman Earle, W. R. Sandbach, William Earle, William Rathbone, and Mr. (afterwards Sir John) Pender, besides a number of other leading merchants of Liverpool and Manchester. Amongst the shareholders were also Michael Bass, M.P. (subsequently Lord Burton), Sir James

[1] Full details of these works appeared from time to time in the public press. Quite recently, when opening up the main streets in Manchester, some of the old iron troughing originally put down was found in excellent condition.

Carmichael, Bart., and George Wythes, of Reigate, a well-known railway contractor. Some of these became directors, as did also Mr. John Watkins Brett, who had become so deeply interested in telegraphic enterprise generally, and more especially submarine telegraphy; and Alderman Watkins, who had been twice Mayor of Manchester. He was one of the old-fashioned sort, and was the last "on change" to stick to a broad-brimmed hat, over leggings and gaiters : whence he was affectionately nicknamed "Old Leather Breeches."

The headquarters of the new Company were located in Liverpool, where most of the capital was represented.

In 1852, the subject of this memoir, when scarcely twenty years of age, was asked by the Board, at the instigation of Mr. Crosbie and Mr. Harrison of the Managing Committee, to become their engineer-in-chief, which post he accepted, resigning his position on the "British." Edward Bright had been manager of the company for some months previously.

It was in this year that the brothers took out their famous patent, to which allusion has already been made. It contained twenty-four distinct inventions connected with telegraphs, and it may be well here to enumerate some of the most important.

First of all, there was the porcelain insulator for fixing aerial telegraph wires mounted on posts.

This has been found to be a highly efficient method of insulation.[1]   It was at once adopted on an

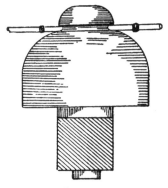

BRIGHT'S INSULATOR

extensive scale, and it continues in use to the present day in one form or another.   There was

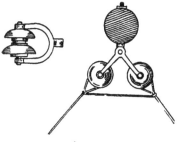

BRIGHT'S SHACKLES

also its adjunct, the shackle or terminal insulator. This is also made of porcelain, and is universally

[1] In his article on the "Electric Telegraph" in the *Encyclopædia Brittanica*, 8th edition, Lord Kelvin, F.R.S., referred to this as "the best idea for a single telegraphic insulator."

employed now, as then, for terminations, and whenever the wire has to be taken at an angle—over houses for instance, round a corner, or in any case where great strains are involved, whether owing to long spans or for other reasons. Then followed the now universal system of aërial telegraph posts with varying length of arms, to avoid the chance of one wire dropping on another.

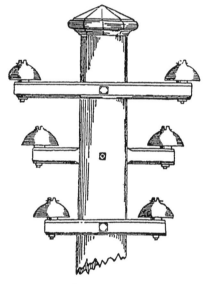

BRIGHT'S TELEGRAPH POST

Next we have the vacuum—or rather, rarefied air—lightning protector, for guarding telegraph lines and apparatus. This has since been—unwittingly, of course—re-patented several times in various forms.

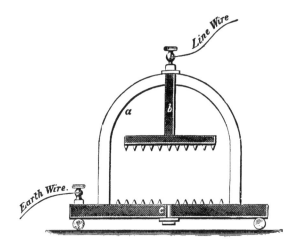

BRIGHT'S VACUUM LIGHTNING GUARD

a Glass chamber, from which the air has been partially exhausted.
b Metallic discharger from line wire.
c Metallic base of protector, connected with the earth.

After this came the brass tape device for the protection of insulated conductors of subterranean or submarine cables. This also has unwittingly formed the subject of later patents.

There was then a translator, or repeater, for re-transmitting electric currents of either kind *in both directions* on a single wire. This contrivance was used with great success by the "Magnetic" Company throughout its system, and was the first of its kind.

Another important item in this famous master patent, was the plan of testing insulated conductors for purposes of fault localisation. This has already been referred to, and a detailed description will be found in the full specification at the end of the volume.

Finally, this patent further comprised a standard galvanometer (foreshadowing differential testing), a new type-printing instrument, as well as what was then a novel mode of laying underground wires in troughs. There were also several other items of less importance, all enumerated in the full specification. The importance of this patent is again testified to by the fact that it is now out of print at the Patent Office.

This patent was taken out when the patentees were respectively twenty-one and twenty years of age; but, as already mentioned, it contained the results of four years' combined thought, part of the drawings being cut out of their old invention book, for Messrs. Carpmael, their patent agents.

In addition to the labour and experiments connected with the practical application of these improvements for the "Magnetic" Company during 1852, Charles Bright directed the completion of a vast telegraphic system throughout the United Kingdom, which had lately been commenced by the Company. It included a main trunk line along the high roads, consisting of ten gutta-percha covered wires, manufactured by the Gutta Percha Company, of Wharf Road, London, laid in troughs underground between London, Birmingham and Manchester, thence by railway to Liverpool and Preston, and six wires onwards, also underground, to Carlisle, Dumfries, Glasgow, and Greenock. From Dumfries a branch of six underground wires was laid under the roads to Portpatrick, to meet the Company's Irish cable. In Ireland, the underground system was extended from Donaghadee to Belfast, and thence, *viâ* Newry and Dundalk, to Dublin, comprising in all nearly 7,000 miles of wire. Although gutta-percha had been discovered in 1843, and its insulating qualities had been appreciated by Faraday and Werner Siemens as early as 1847, this was the first instance, in our country, in which any length of gutta-percha covered cable had been laid underground. Siemens was probably the first to cover cables in this manner, and his example was soon followed by Mr. Bewley, of the Gutta-Percha Company. Some lines in Germany, laid down by Siemens and Halske for the Prussian Government, preceded those of the " Magnetic," who in their turn were followed two years later by the " Electric " Company, who adopted gutta-percha as a covering for its underground system of cables.

It would be well here to consider the *nature* of this underground system. The form it should take was very carefully considered by Charles Bright.

It was evident that the integrity of the insulating coatings of gutta percha could not be preserved long without some external protection throughout the length of each line, as the mere compression of the soil, gravel and stones would at once have injured it; and in opening the roads for repair they would experience still further damage.

After discussing the merits of various plans of protection, it was finally decided that the wires throughout towns should be laid in 2½-inch cast-iron piping, divided longitudinally so that the wires might be laid in quickly without the tedious and injurious operation of drawing them through, as was the case with the old system of street work, where the wires were laid in ordinary gas-piping.

On the other hand, Bright decided that along the country roads, which were comparatively little liable to disturbance from the construction of sewers, or laying of gas or water pipes, the wires should be laid in *creosoted* wooden troughs of about 3-inch scantling, cut in long lengths, so as to be almost free from the chances of damage upon any partial subsidence of the soil, which not unfrequently occurs in districts where mining operations are carried on.

The tops of the troughs were to be protected by fastening to them a galvanized iron lid.

Some idea of the trough system for the public highways may be gathered from the accompanying sketch.

The gutta-percha-covered wires were deposited in the square creosoted wooden trough (shown below), after being bound together by a lapping of tarred yarn. To deposit the rope of insulated conductors in the trough it was first coiled upon a large drum, and this was then rolled slowly over the trench, which had a depth of some three feet. The rope of wires was paid off easily and evenly into its bed. A galvanised iron lid, about an eighth of an inch thick, was then fastened on by

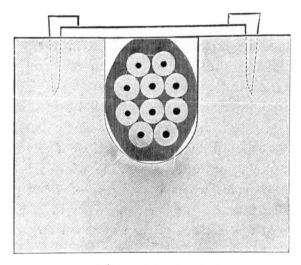

BRIGHT'S UNDERGROUND SYSTEM

clamps (see illustration), and the trench filled in again. The whole system is fully described in Shaffner's *Telegraph Manual* (New York), and also in his *Telegraph Companion.*[1]

[1] Some time previous to the publication of the latter Mr. T. P. Shaffner addressed to Charles and Edward Bright a number of questions with reference to the European System of Telegraphs, which were duly answered. Both questions and answers are given in the Appendices to this volume.

LAND TELEGRAPHS

The method adopted in the case of underground wires laid in iron troughs under the streets of towns has been already described in the case of the work done by Charles Bright at Manchester for the " British " Company a year previously.

This new undertaking was carried out by the Company's staff, assisted by the late Mr. W. T. Henley and Messrs. Reid Brothers, the famous contractors of that time.

The following letter, addressed by Charles Bright to his *fiancée*, may be of interest here, as picturing the scene :—

Manchester, *September* 11*th*, 1852.

Your letter did not arrive until last evening. I should have written in advance to send, but I have been very busy. Last night I spent entirely out of doors, and as I have not been able to get any sleep since, I shall not write long now. . . . It is the third bedless night I have had lately, and I expect two more next week.

I was at Liverpool last night, getting our wires from the station to our offices in the Exchange. From the great traffic during the day, it is impossible either in Liverpool or Manchester to do anything by *day*, and unless I keep a sharp eye on the men, either the pipes are laid too near the surface, or they break gas or water pipes and cause expensive repairs. Moreover, they never do a third of the work at night unless I am with them !

Last night I did the quickest piece of telegraphic work which has ever been done. We began at ten, and by eight in the morning we had laid piping containing eight wires under the streets nearly half a mile, and all re-paved.

Can you fancy such a scene ? A long row of men with

pickaxes, followed by others with spades, and after them a gang of men laying pipes and wires and, to conclude, another set re-laying the paving stone. This row of workmen are lighted up by large fire-grates at intervals, flaring and smoking away like beacons on the coast—a perfect Babel of voices—the continual sharp knocking of the pickaxes and the scraping and clanging of the pipes being laid, and hammered up, added to continued shouting for this or that tool. If you can conjure up this, you can fancy my long figure appearing in the light here and there with two or three foremen—quite in my element, only I don't like the night. I expect you would be very much alarmed if you unexpectedly were awoke by such a noise and looked out on such a scene! . . .

I tell you all about my night's doings, because I was pleased at the speed which I had calculated before on doing it in. The plan was a new one of my own. . . .

One of Bright's assistants has described how his chief wrote out instructions to the minutest details even to the extent of stating where the vessels of pitch were to be placed, besides specifying the temperature of the mixture, and that it was to be tested before being run into the trough.

The great advantage gained by laying these main trunk lines underground was that they were thereby absolutely beyond the reach of damage by stormy weather. Thus, it was that the "Magnetic" Company became at once a prosperous and successful company.

Charles Bright also personally directed the erection of overhead wires on the following railways : The East Lancashire, Caledonian, Midland, Great Western, Great Southern and Western, Waterford and Limerick, Dublin and Drogheda, Belfast Junction, Ulster, County Down, Belfast and Coleraine, Londonderry and Enniskillen, Londonderry and Coleraine.

The Journal of the Institution of Civil Engineers, in its obituary notice,[1] contains the following testimony : "All this work, both overhead and underground, entailed a vast amount of energy and perseverance on the part of Sir Charles Bright, and many are the stories related of the difficulties overcome in the rapid progress of the underground work."

The summer of 1853 saw great events in Charles Bright's life. He married, at the age of twenty-one, Miss Taylor, daughter of Mr. John Taylor, of Bellevue, Kingston-upon-Hull, to whom he had for some time been devotedly attached. Mr. Taylor was head of one of the leading mercantile firms in Hull. The Taylors, and their ancestors the Willots and the Gills, also came from the West Riding of Yorkshire originally—*i.e.* from Treaton, Chesterfield, etc. Curiously enough there had been, in former generations, an intermarriage between the two families, for Paul Bright of Inkersall married a Taylor—not to mention other "attachments." Charles Bright's *fiancée* was one of

[1] *Mins. Proc. Inst. C.E.*, vol. xciii. part iii.

the youngest in a family of nine. The young couple had become engaged nearly two years previously. They had met first whilst staying with mutual cousins, the Henry Brights, near Hull. In that way Charles saw much of the Taylor family in their home life. The wedding took place on May 11th, 1853, at St. James', Hull. They started life together on an income of about £250, and have often looked back with pleasure on those early days of comparative poverty which were, nevertheless, some of the happiest of their life together.

In this year (1853) the first effective cable to Ireland, was made, under Bright's supervision, by Messrs. Newall & Company, of Gateshead,[1] and laid between Donaghadee, in Ireland, and Portpatrick, in Scotland.

At the outset of the "Magnetic" Company's operations, the brothers found it necessary to devise fresh apparatus to compensate for the inductive discharge resulting from the long underground circuits, by discharging to earth and thus neutralising the recoil currents. From that time till the spring of 1854, they carried out a series of experiments on the great lengths of subterranean wires under their control, in order to investigate this novel phenomena presented by the inductive effects

[1] A firm of contractors who had been largely responsible for the English Channel cable of 1851, and who had subsequently manufactured, and, in some cases laid, other pioneer lines.

MISS TAYLOR
(afterwards Lady Bright, from a drawing by her youngest daughter)

61

of the long gutta-percha-covered conductors, with the view to working through an Atlantic cable. This had been the great object which Charles Bright had in view in pushing on the Company's extension in the West of Ireland, his idea being at the time, that a point between Limerick and Galway would be the most suitable landing-place for the cable. Some of the results of these experiments were detailed and illustrated experimentally by Mr. Edward Bright, at a meeting of the British Association at Liverpool, in 1854,[1] in an address on " The Retardation of Electricity through Long Subterranean Wires," for which he received the appreciative thanks of the President of the " Section."

During 1854, the brothers were heavily burdened—Charles in completing the enormous network of telegraphic wires—thousands of miles in all—that had been constructed under his direction with such wonderful rapidity throughout the kingdom; and Edward in acquiring and fitting up the stations, organising the staff, making rules and regulations for the service, arranging message tariffs, and supply of news to the press, etc.

Time was nevertheless found for other work. They engaged in experiments with the late Mr. Staite, on the electric light—then absolutely in its infancy. Mr. Staite's arc lamp had been

---

[1] See *B. A. Reports*, 1854.

exhibited for some months on the Liverpool landing-stage, till the pilots complained (as well as the steamboat captains) that it dazzled them and hindered their steering on the river Mersey. Mr. Staite—whose share in the early development of electric lighting has been dilated upon by his son—was a neighbour of Edward Bright, in Cheshire; and, as dynamo machines for electric lighting were not then practically developed, the brothers were able to greatly assist his temporary installation, with the very powerful electric batteries which they employed in their own experimental researches.

Charles also devoted some time to a series of interesting investigations in connection with a dynamo machine which had been devised by Mr. Soren Hjorth, and shown by him at the Liverpool Dispensary. Powerful dynamic currents were produced by this apparatus which, though not economical, have considerable historic interest.

At this period, both the brothers materially aided the late Admiral Fitzroy in the inauguration of his plan of daily telegraphic reports in connection with the newly-born Meteorological Department of the Board of Trade, and the storm-warning system which the Admiral had organised. They arranged for the requisite barometers, thermometers, wind gauges, etc., to be set up at a number of the Magnetic Company's stations, especially in the West of Ireland and Scotland, including Cape

Clear, Limerick, Tralee, Galway, Portrush, Ayr, Androssan, etc., where coming changes of weather are, as a rule, first indicated from the Atlantic. The "Magnetic" staff were duly instructed in the taking of observations twice daily. These were then telegraphed—as now, though less fully—to the Meteorological Office in London by means of a concise code, drawn up by Admiral Fitzroy and Charles Bright, with a view to expediting the messages by reducing their length, as it did, by about one-fifth. For this work the formal thanks of the authorities was communicated in the Astronomer Royal's report of that year. Although at the outset these weather forecasts were somewhat tentative and were much scoffed at— and although many shipowners charged them with being the cause of keeping their ships in port unnecessarily—their vast utility in lessening the danger to life at sea was not long in being recognised, and the forecasting of the weather has now, by dint of experience, become almost an exact science.

The brothers also busied themselves in bringing the correct time into Liverpool—where shipping people, indeed everybody, much needed to know it—from the Observatory, which was then under the control of Mr. J. C. Hartnop, F.R.A.S. In doing so a novel system, invented by Mr. R. L. Jones, of Chester, was adopted. This did not attempt to *drive* public clocks by very strong currents, but only to *bridle* their pendulums. This ingenious contrivance was applied to the heavy, not to say ancient, clock in the Town Hall at Liverpool—previously a notorious fugitive of real time—and kept it accurate to the second thereafter without trouble. Jones' system

BRIGHT'S MAGNETO-ELECTRIC TELEGRAPH
(*Nicknamed " The Thunder-Pump"*)

was also applied to the large clock outside the "Magnetic" offices, as well as to many others.

At the end of the year 1855 a revolution was effected in the telegraphic apparatus used by the "Magnetic." This company, originally formed to work Mr. W. T. Henley's magneto-electric telegraph instruments, had up to the period in question continued to employ that class of apparatus, with important improvements introduced by the Brights. The magneto-electric currents were produced by the operator who, by means of a handle, caused magneto-electric coils to move on an axle, to and fro, before the poles of permanent magnets. This machine was appropriately nicknamed a " thunder-pump." The currents so produced moved magnetic needles to and fro at the further end of the line, the combinations signifying the different letters, as with the Voltaic telegraph of Cooke and Wheatstone already referred to.[1]

Charles Bright, however, perceiving the objection to any instrument based on visual signalling, set to work to devise an apparatus which would communicate signals to the ears. The result was that in 1854 he produced the Acoustic Telegraph, since commonly known as " Bright's Bells."

This invention is described fully in the Patent Specification, No. 2,103, of 1855, reproduced in the Appendices, Vol. II., and also by Noad and

[1] For a full description, see *The Student's Text Book of Electricity*, by H. M. Noad, Ph.D., F.R.S., revised by W. H. Preece, C.B., F.R.S. (London : 1895, Crosby Lockwood, & Son). Also the *Mechanics' Magazine*, July 23, 1853 ; as well as the Patent Specification No. 14,331 of 1852 reproduced in full at the end of Vol. II. (see Appendices).

Preece, in their previously mentioned text-book, in the following terms :

"Under the ordinary system of telegraphing, it is necessary to employ a transcriber to write down the words as interpreted from the visual signals and dictated to him by the receiving operator, whose eyes being fixed on the rapidly moving needles could not be engaged in conjunction with his hands in writing. It was found that, owing to the frequent occurrence of words *of nearly similar sound*, the transcriber sometimes unavoidably misunderstood the meaning of the receiving operator, and altered the sense of the despatch by writing the wrong word. Such words as two, too, to ; four, for ; hour, our, etc., may, for instance, be very easily confounded. These errors cannot, however, arise when the clerk, who, having heard each word pass through the acoustic telegraph *letter by letter*, is able, his eyes being at liberty, to himself write what he has received without the aid of an amanuensis. Besides the saving in staff (of writers) and in mistakes, any injury to the eyes of the clerks is prevented, and an appeal is made to an organ far better capable of endurance and accurate interpretation." [1]

[1] Another excellent description was given by Lord Kelvin, F.R.S. (then Professor Thomson), in his article on the "Electric Telegraph" in the eighth edition of the *Encyclopædia Brittanica*, vol. xxi. (Edinburgh: A. & C. Black).

The general principle of this instrument consists in the sounding of two bells of different
pitch by different currents. The letters and
words are readily formed from the difference in
their tone and the number of beats, the same
(Morse) alphabet being employed as in other
telegraph systems.

The nature of the apparatus is shown in the
accompanying illustration :

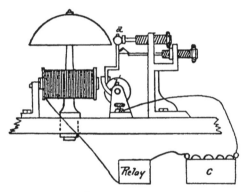

BRIGHT'S BELL INSTRUMENT.

a is the hammer of the bell, held back to a
stop by a flexible spring. The rod of the
hammer is fixed to the projecting horns of the
movable soft iron core of an electro-magnet *b'*.
This electro-magnet *b'* is placed opposite to a
fixed horse-shoe electro-magnet *b* ; and the connections are so arranged, that, on the current
passing from the relay, the electro-magnets are
polarised with their opposite poles to one

another. Upon a current passing, the bell affected is at once struck, and the bell being muffled so as to produce a short sound, the blow may be repeated as rapidly as desired without any vibration caused by one sound interfering with that succeeding it.

A local battery supplies the mechanical power required to strike the bells. The battery is put in connection with either bell according to the current—positive or negative—passed through a relay, shown in the next illustration, where also may be seen the general arrangement.

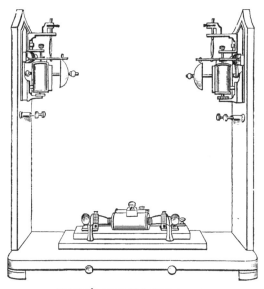

BRIGHT'S ACOUSTIC TELEGRAPH

Here the receiving clerk, with his head between the two different toned bells, each fixed to a wooden partition, can readily distinguish the signals corresponding to the beats of the needle. As fast as he does so he writes down their significance.

The keys with which currents are sent to work this apparatus are of a simple commutating form. By pressing down one key, the current is made to pass in one direction, and in the reverse when the other key is used.

This form of telegraph, like the Morse (sounder or writer) and other instruments of to-day, requires one wire. In point of speed, however, it has a great advantage, as it utilises both positive and negative currents, while the Morse is only available for one current. Thus, the acoustic instrument only occupies in the transmission of the alphabet about half the time of the American apparatus, and is, moreover, much faster [1] than any type of visual telegraph (except, of course, those worked on the Wheatstone automatic system), for reasons already explained, besides being much more accurate.

This invention of the Brights is still largely used, after a lapse of nearly forty-five years. It is so simple in its working, and so much appreciated by the telegraphic employées, that the number of these instruments now in use by the Government telegraphs is nearly five times

[1] A speed of forty words a minute is frequently attained.

as great as that employed by the "Magnetic" Company at the time of the transfer of the telegraphs to the State, in 1870. As it is still one of the fastest instruments extant, this number is being constantly added to, mainly owing to the large increase in the number of press messages. This apparatus was adopted on all the "Magnetic" Company's lines as soon as it was perfected, in 1855.

In that year young Bright thought out another important invention with his brother. This consisted of a system of duplex telegraphy, fully illustrated and described in the same specification (see Appendices.). This was worked successfully between London and Birmingham. As, however, the "Magnetic" Company's traffic did not then fill their wires, the system was temporarily laid on one side.

During the year 1856 some of the Magnetic Company's underground lines began to give trouble. This was largely caused by the decay of the gutta-percha, which, laid in a dry soil, lost much of its essential oil, by evaporation : its woody substance only remained—and this in a porous condition.

To overcome the above difficulty Charles Bright devised an apparatus for first forcing a liquid insulating substance into the pores of the decayed gutta-percha, and then giving it an outer coating of a more solid material. The plan, however, was not found sufficiently satisfactory in practice. The authorities thereupon set themselves to consider how they could best extend their overhead system. This culminated in the absorption of the British Telegraph Company, which had exclusive rights (as before mentioned) for overhead telegraphs along the public roadways. There was no difficulty in arranging the matter, for, though the British Company had a few years earlier combined with the European and American Telegraph Company[1] in connection with the Submarine Company's cables, they had been unable to pay any dividend to speak of.

After the above amalgamation, under the title of the British and Irish Magnetic Telegraph Company—still familiarly known as the "Magnetic"—the underground wires were only used in places where circumstances rendered them specially desirable. It is interesting to note, however, that during recent years there has been a return to the subterranean system, especially in Germany and France, owing to the fact that wires can be less readily cut when underground, in time of war, and for the original reason which guided the Brights,—*i.e.* comparative shelter from the weather. The new "Magnetic" had an agreement with the Submarine Telegraph Company,

---

[1] Formed by the Bretts, for working the House Printing Telegraph.

under which the whole of the latter's cables were to be worked in connection with the land lines belonging to the former. Charles Bright remained engineer-in-chief to the Magnetic Company until about 1860, from which time (owing to press of other work) he held a consulting position only. Thereupon Edward Bright assumed the engineership—with Mr. R. Dodwell, formerly district engineer to the Electric Company, as his assistant engineer—in addition to the general management; and these joint positions he held until the transfer of the Company's business to the Government in 1870, and the consequent winding up of the Company itself.

In connection with the "Magnetic," the London District Telegraph Company[1] was promoted, in 1859, for the purpose of establishing 100 stations in and about London, and for working on a cheap sixpenny tariff for twenty words—which was, however, found to be unremunerative—all provincial and continental messages being transferred *viâ* the "Magnetic" *réseau*. The London District lines were entirely on the overhead system. Charles Bright was appointed consulting engineer to this Company, and he subsequently accepted a seat on the Board. Mr. C. Curtoys was the Company's manager and secretary.

---

[1] Afterwards the London and Provincial Telegraph Company.

LAND TELEGRAPHS

In order to give a fairly complete history of these early land telegraphs, we must finally say a few words with regard to the other companies contemporaneous with those in which Charles Bright was concerned.

The United Kingdom Telegraph Company was formed, in 1858, to work Allan's needle instrument. This system was an overhead one, and ran through towns. In the country it had a right of way along the route of canals. The concern has an historic interest on account of its being the first to start a system of universal shilling messages — 20 words and address free. This scheme, however, appears to have been before its time, for it was not altogether a success. Mr. Thomas Allan (afterwards of light cable fame) was the engineer of the Company, whilst they had a very energetic secretary in Mr. William Andrews, an old friend of Charles Bright's. Mr. Andrews is now chairman of several of the big cable-working companies, and has been for the last twenty-seven years managing director of the " Indo-European."

Other telegraph companies also there were, and which earned good profits, such as Reuter's, and the London and Provincial. The business, indeed, of some of those with which Charles Bright was connected flourished so much that they were able to pay dividends as high as 15 per cent. per annum, the Magnetic Company maintaining a steady dividend of not less than 12 per cent. for many years.

Amongst the friendly contemporaries of Charles Bright in the early land telegraphs were Mr. Edwin Clark, engineer to the Electric Company, and his brother, Mr. Latimer Clark, who afterwards filled the same post, followed in his turn by Mr. R. S. Culley. There were also Mr. C. F. Varley and Mr. F. C. Webb, the former electrician to the Electric Company, and the latter one of its engineers, whilst Mr. W. H. Hatcher and Mr. Henry Weaver acted as engineer and secretary respectively at different periods. Amongst the survivors of early telegraphy in this country there are also Mr. C. E. Spagnoletti, Mr. W. T. Ansell (both of the old "Electric"), and Mr. Richard Collett, Mr. Charles Gerhardi and Mr. James Gutteres — all of the "Magnetic." All of these were close friends of Charles Bright's and were—as we shall see—associated with him subsequently in other telegraphic work.

The Telegraph Act for the settlement of terms with the companies was passed in 1868. The following year the Telegraph Purchase and Regulations Act (for the administration of Government service) became law. Up to this time our Government was the only one, besides that of the United States, which had not undertaken the control of the country's telegraphic system When at last the transfer took place,[1] the companies had held the field for twenty-four years.

[1] For further particulars regarding the early telegraphs of this country, and with reference to the present system under State control, the reader is referred to a series of articles in vol. xvi. of the *Electrical Review*, entitled "The Telegraphs of the United Kingdom," by Charles Bright, F.R.S.E. (1897).

Chapter IV

THE CABLE TO IRELAND

A T the date of the first cable to Ireland, two
submarine cables had already been submerged. The first serious attempt was that projected and primarily promoted by the brothers
Brett : this was eventually carried to a successful
issue in 1851, by Mr. Thomas Russell Crampton, a
civil engineer of distinction. Prior to these, in 1849,
an experimental line with a gutta-percha core had
been laid by Mr. C. V. Walker, F.R.S., in the
English Channel, for some distance off Folkestone.
Also in the following year, another unprotected
gutta-percha insulated conductor had been laid
between England and France by Mr. Charlton
Wollaston, acting as engineer to the Submarine
Telegraph Company. Though the latter failed
—through want of armoured protection—to be
effective, an unarmoured core was laid four or five
years later in the Black Sea, which, owing to the
nature of the bottom in which it lay, had a somewhat longer term of useful existence.

The second, line was that between Dover and

Ostend, being also on behalf of the Submarine Telegraph Company.[1]

Thus, this was the third submarine cable communication successfully carried out. It was, however, in much deeper water than had hitherto been

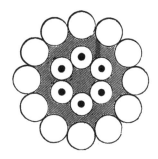

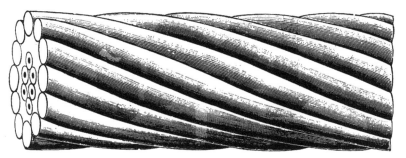

THE ANGLO-IRISH CABLE, 1853

[1] Since Bright's death this Company's cable system has been absorbed by the State and worked by H.M. Post Office. The transfer of the business took place in 1889, and has proved a serious matter pecuniarily to Sir Charles' family. The late Sir Julian Goldsmid, as chairman, did his best to bring about a satisfactory arrangement ; but, in the end, the Company and its shareholders came off very poorly at the hands of the Government.

experienced. As three previous attempts (made by others) to lay a line across the Irish Channel had failed,[1] every care was taken by the Company to ensure success.

An important improvement was made in the design of this cable as compared with what immediately preceded it. In this case an inner bedding of yarn was supplied for the six insulated wires (see illustration). The total weight of this cable was seven tons to the mile.

The manufacture of the line was carried out unaccompanied by any serious mishap. As fast as it was made, it was coiled up on the wharf ready for shipment.

When the time for shipment came, the massive six-core cable was stowed away in the hold of the cable-laying vessel in an oblong coil.

It so happened that the laying of this line had to take place during the days closely following upon Charles Bright's marriage. The expedition was graced by the presence of his bride, who was thus able to assist at the telegraphic union of Great Britain and Ireland. This, so far as we are aware, is still the only occasion on which a lady has been aboard during actual cable-laying operations.

[1] One of these being very light, owing to having no iron armour, was literally carried away and broken to pieces by strong tidal currents.

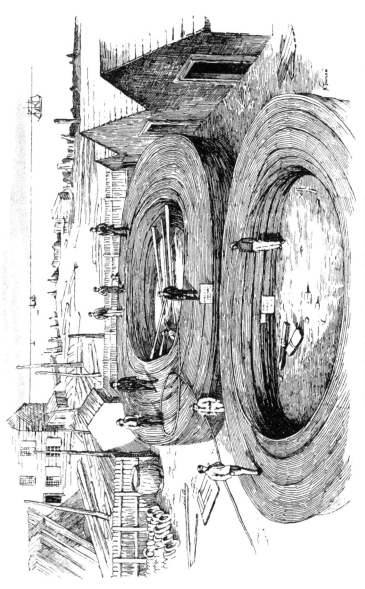

COILING THE CABLE READY FOR SHIPMENT

THE CABLE TO IRELAND

The expedition consisted of the screw steamer *William Hutt*, named after a former M.P. for Hull (with the cable and apparatus on board), the *Conqueror*, and the *Wizard*. The ships were under the navigating control of Captain Hawes, R.N., specially appointed by the Admiralty ; and this officer rendered invaluable assistance in determining the exact course to be taken without which the squadron could probably not have overcome the swift tides and adverse currents associated with that part of the Irish Channel.

Besides Charles Bright and his bride there were on board during this expedition, his brother (as manager of the Company), and Mr. Newall,[1] the contractor ; also Mr. Statham, of the Gutta Percha Company, Mr. William Reid, and Mr. T. B. Moseley.

Starting operations from the Irish coast, the shore end of the cable was first landed at a point about two miles from the south of Donaghadee harbour, Co. Down, and the laying of the deep sea cable was then proceeded with.

This undertaking was not, however, without its vicissitudes. The arrangements and apparatus then employed for submerging a cable were, it need scarcely be said, not of the complete character with which experience has endowed us at

[1] A partner of this gentlemen was Mr. Charles Liddel, whose family were connected with the Brights in former generations : See Genealogical Tree (Appendices).

the present day. No central guiding cone or cylindrical frame were in use at that time, nor yet any "crinoline." Each coil was turned bodily over by the men below to take the turn out in emerging to the guide pulley above, whence it passed through a rotometer or speed measurer to a large drum on deck. Round this drum it took several turns before passing into the sea over an

THE PAYING-OUT APPARATUS

iron rail at the stern. The drum was fitted with a flexible iron strap on its circumference, attached to a lever hand-brake, to check the cable's rate of delivery outboard. Without this precaution in the deeper water (nearly a quarter of a mile in places) the heavy monster would have "taken charge" altogether.

As it was—when a heavyish sea arose about midway across—notwithstanding the efforts of the

THE IRISH LANDING-PLACE

men in the hold, one turn got on several occasions under another, making a "foul flake," which would pass up in a tangled mass. This necessitated the stopping of the ship and a temporary cessation of paying-out operations till the great knot was unravelled. Such an operation as this is no easy matter when the extreme rigidity of this heavily armoured cable, with its twelve stout iron wires, is considered.

Thus it was that the expedition did not arrive and anchor off Port Patrick, on the southern border of Wigtonshire, until midnight, the landing of the shore end being deferred till the following morning.

At Port Patrick, Mr. Robert Crosbie (the Chairman of the Managing Committee), who had been one of the leading promoters of the Magnetic Company, with Mr. Henry Harrison, a most (active director), were awaiting the landing of the cable.

This final operation was performed, amid much enthusiasm, in Mora Bay, a little to the north of Port Patrick, belonging to Mr. Blair, of Dunskey, who kindly awarded permission to lay the land wires through his estate. As soon as the cable end had been taken up to the position assigned for it, the signalling apparatus was put into operation, and the following message despatched to Dublin :—

SIR CHARLES TILSTON BRIGHT

"Mora Bay, Port Patrick,

"*May* 23, 1853.

"The Directors of the English and Irish Magnetic Telegraph Company beg to acquaint His Excellency the Lord Lieutenant that they have this morning successfully effected communication between the shores of Great Britain and Ireland by means of a submarine cable from Port Patrick to Donaghadee."

This cable lasted, with slight repairs, for many years—up to, and long after, the purchase of the Magnetic Company's lines by Government, in 1870.

In later years, referring to this expedition, Sir Charles Bright used often to humorously remark that, so long as we had telegraphic communication with Ireland, there could be no possible need for discussing the question of Irish Home Rule.

LAYING THE CABLE TO IRELAND, 1853

Chapter V

THE ATLANTIC CABLE

Section I

Investigations and Stepping-Stones

WE have now come to the most arduous, besides the most interesting and memorable achievement of Charles Bright's career, namely, the telegraphic linking of England and America by submarine cable.

In *A Midsummer Night's Dream*, Shakespeare makes Puck say, " I'll put a girdle round about the earth in twenty minutes!" Though little Puck never carried out his boast, the subject of our memoir in the undertaking here referred to went some way towards realising it in practice.

From this he acquired such fame, whilst only twenty-six years old, as few men engaged in carrying out the great works of the world, can ever hope to attain. This achievement was characterised by the *Times* newspaper as "the accomplishment of the age," and by Prof. Morse as "the great feat of the century."

The part Bright took in this then unprecedented enterprise included the scientific demonstration of its practicability, the projection, the provision of capital, the organisation, and the ultimate successful laying of 2,200 miles of cable across ocean depths of two to three miles, in the face of storms, repeated breakages, and every kind of difficulty.

By his scientific knowledge, ingenuity, and determined pluck, he carried it through at a time when only a few short cables had been successfully laid —mostly in comparatively shallow water [1] — and

[1] First of these was the English Channel line (in some thirty fathoms of water), already referred to briefly in connection with the Anglo-Irish cable laid by Charles Bright in 1853. In the same year direct submarine communication had also been established with Holland and Belgium—due mainly to the efforts of the late Mr. Edwin Clark and Mr. F. C. Webb, working on behalf of the Electric and International Telegraph Company—as well as between Denmark and Sweden. Next came the projection of various telegraph systems under the Mediterranean, and much difficulty was experienced in the case of these latter, which were in turn eventually laid under the auspices of Mr. J. W Brett and Mr. R. S. Newall. There was also the Black Sea (bare-core) line, carried out temporarily by Messrs. Newall & Co., for use during the Crimean War.

But nothing so daring as a submarine ocean cable, laid in an open seaway, had yet been attempted, and in his Presidential Address to the Institution of Electrical Engineers, in 1889, Lord Kelvin, referring to this work, said : " We must always feel indebted to Sir Charles Bright as the pioneer in that great work, when other engineers when the art of submarine cable work was in its infancy as regards construction, insulation, and mechanical appliances.

Many at Bright's age would have flinched at the responsibility with so limited an experience.

It scarcely signifies who expressed the first crude notion of the possibility of trans-Atlantic telegraphy. Some theoretical scientists were so venturesome as to predict ocean cables several years before gutta percha had been introduced as an insulator, and still longer before even the shortest cable had been made. And it is interesting to note how speculative minds were at once attracted by the notion of a line uniting the two great English-speaking nations.

The earliest record of practical signalling under water of which there appears to be any detail[1] relates to experiments made by Colonel (afterwards Major-General Sir F. C.) Pasley, of the Royal Engineers, in 1838, in the Medway at Chatham, and again in connection with the submarine operations on the sunken wreck of the *Royal George*, off Spithead. In these first cases the conducting wire was bound round with strands of tarred rope, and then again with pitched yarn.

Early in 1839 Dr. O'Shaughnessy (afterwards

would not look at it, and thought it was absolutely impracticable."

[1] *Submarine Telegraphs*, by Charles Bright, F.R.S.E. (London : Crosby Lockwood & Son).

SIR CHARLES TILSTON BRIGHT

Sir William O'Shaughnessy Brooke, F.R.S., Director of the East India Company's Telegraphs) laid a subaqueous telegraph across the broad river Hooghly, about one and a half miles in length, which he thus describes in a paper contributed to the Journal of the Asiatic Society, September, 1839 : " Insulation, according to my experiments, is best accomplished by enclosing the wire (previously pitched) in a split rattan, and then paying the rattan round with tarred yarn ; or the wire may, as in some experiments made by Colonel Pasley at Chatham, be surrounded by strands of tarred rope, and this again by pitched yarn. An insulated rope of the above description may be spread along a wet field, nay, even led through a river, and will still conduct (without any appreciable loss) the electrical signals above described." [1]

In 1840 Professor Wheatstone brought forward designs for making a cable, to be insulated and protected by tarred and pitched yarns similar to Colonel Pasley's method, and went into many details respecting the machinery and appliances for its manufacture. In 1843 Professor Morse wrote to the Secretary of the American Treasury, reporting some trials in New York Harbour, and ending his report with the following prophetic words : " A

[1] A very full account of O'Shaughnessy's work was given by Mr. P. V. Luke, C.I.E., in the course of an original communication, in 1891, to the Institution of Electrical Engineers.—*Jour. I.E.E.*, vol. xx. p. 102.

telegraphic communication on the electro-magnetic plan may, with certainty, be established across the Atlantic Ocean! Startling as it may now seem, I am confident the time will come when this project will be realised."

It was on the 16th of June, 1845, that Mr. John Watkins Brett registered the General Oceanic Telegraphic Company, the object specified being: " To form a connecting mode of communication, by telegraphic means, from the British Islands, and across the Atlantic Ocean to Nova Scotia, and the Canadas, the Colonies, and Continental Kingdoms."

But these projected works, and many others about the same time, were necessarily merely tentative, for, as we have already explained, the essential insulating medium—namely, gutta percha—was not introduced as an insulator till 1847; and until then there did not exist the means of satisfactorily insulating, nor had suitable apparatus been designed, for the working of submarine cables. The question, therefore, is not as to who first had the crude notion, but who put it into a practical form and overcame the difficulties.

Before the Atlantic Telegraph could assume a practical shape, the following essential requirements had to be dealt with :—

1. Ocean soundings, showing the depths and nature of the sea-bottom, had to be taken and placed on record.

2. Experiments had to be made to prove that a con-

ductor, insulated with gutta percha, and of the necessary length (about 2,000 miles), could be signalled through for telegraphic purposes.

3. A suitable form of cable for the specific purpose had to be designed.

4. Provision had to be made to prevent competition, so that—for some time, at least—a fair return might accrue to those who staked their capital in what then appeared so risky an enterprise.

5. The confidence of the moneyed mercantile class, who who would mostly benefit by such a means of communication, had to be won ; and

6. Government recognition had to be obtained, and, if possible, Government subsidies.

Lieut. O. H. Berryman, U.S.N., who was employed in the U.S. brig *Dolphin* to make observations on the winds and currents along the main sailing route to England, had also run a line of deep sea soundings in the Atlantic basin between Newfoundland and Ireland in the summer of 1856, from the U.S. steamer *Arctic*, by means of the ingenious apparatus designed by Lieut. J. M. Brooke, U.S.N., which may be described as follows :—

A light iron rod (*C*) hollowed at its lower end into a thin, specially-made line, and passed loosely through a hole in the centre of a weight, such as a cannon ball (*A*), which is attached to the line by a couple of links. On the bottom being touched, the links reverse position, owing to the weight being taken off, and so set free the cannon ball (*B*) or plummet. The line is then free to be drawn up again with merely the light rod. The latter has had its open cylinder-end pressed into the ground, and is thus drawn up with a specimen of the bottom (unless rock or coral) enclosed. The minute the bottom is touched, the speed of the line running out from its drum is reduced, for the weight no

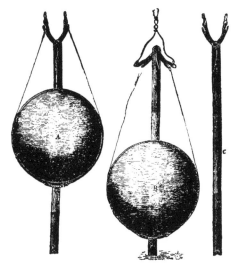

BROOKE'S SOUNDING APPARATUS

longer drags it down ; and so, by means of a revolution-counter, the depth is accurately ascertained.

The soundings gave a general depth of about two miles and a half, gradually shoaling on the Newfoundland side, but rising more quickly towards the Irish shore. The entire proposed route was marked by an oozy bottom, of which specimens brought to the surface were shown under the microscope to consist of the tiny shells of animalculæ—the indestructible outside skeletons of *diatomaceæ* and *foraminiferæ*. These live near the surface of the ocean in myriads upon myriads, incessantly sinking downwards to the bottom as their short life is ended. Thus, in the course of

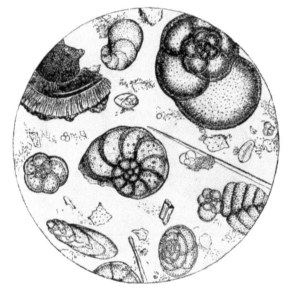

SPECIMEN OF THE BOTTOM OF THE ATLANTIC "TELEGRAPH PLATEAU"
[*Magnified* 17,000 *times*]

ages, there grows constantly upwards a formation similar to the chalk cliffs of England, which contain the identical shells, deposited when this country was submerged far below sea level, many thousands, perhaps millions, of years ago.

No sand or gravel was found on the ocean bed, from which it was deduced that no currents,

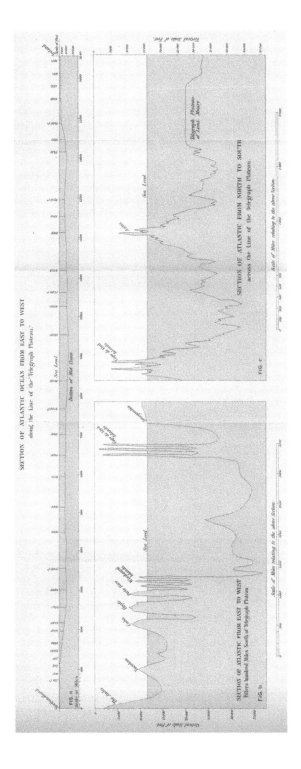

The material originally positioned here is too large for reproduction in this reissue. A PDF can be downloaded from the web address given on page iv of this book, by clicking on 'Resources Available'.

or other disturbing elements, existed at those depths; for otherwise these frail shells—with their sides as thin as any soap or air-bubble—would have been rubbed to pieces. As it was they came up entire, without a sign of abrasion. The plateau, or ridge, which extended for some 400 miles in breadth, was, in fact, a veritable feather-bed for a cable, when once weather and other conditions allowed of its safe submersion.

Lieut. M. F. Maury, U.S.N., Chief of the National U.S. Observatory, to whom the observaations and results of Lieut. Berryman were referred, made a long report to the Secretary of the U.S. Navy, dated February 22, 1854, in which he remarked : " This line of deep-sea soundings seems to be decisive on the question of the practicability of a submarine telegraph between the two continents, *in so far as the bottom of the deep sea is concerned.* From Newfoundland to Ireland the distance between the nearest points is about 1,600 miles ; and at the bottom of the sea between the two places there exists a "plateau," or shallow platform, which seems to have been placed especially for the purpose of holding the wires of a submarine telegraph, and of keeping them out of harm's way.[1]   . . But

[1] Lieutenant Maury's views, as here expressed, were illustrated by the opposite plate. A glance at this gives a general idea of the even surface afforded by that part of the Atlantic bed where the cable was to rest as compared with the Alpine character, which prevails north and south of the "plateau," or platform.

whether it would be better to lead the wires from Newfoundland or Labrador is not now the point at issue ; nor do I pretend to consider the question as to the possibility of finding *a time calm enough, the sea smooth enough, a wire long enough, or a ship big enough* to lay a coil of wire sixteen hundred miles in length. Still, I have no fear but that the enterprise and ingenuity of the age, whenever called upon to solve these problems, will be ready with a satisfactory and practical solution of them."

Similar conclusions to these were arrived at from the soundings taken in the North Atlantic by Commander Joseph Dayman, R.N., in H.M.S. *Cyclops*, a little while later.

The possibility of laying an Atlantic line having taken firm hold in the mind of Charles Bright, ever since the successful laying of the cables to France, Ireland, and Belgium, he and his brother Edward carried on from 1853 to 1855, a series of extensive experiments on the great lengths of underground gutta-percha-covered wires of the Magnetic Telegraph Company, which were under their management. In these wires the conditions were similar, electrically speaking, to those existing in the case of a submarine cable. By linking the wires to and fro between London and Dublin—including the conductors of one of the Irish cables—or employing the ten wires between London and Manchester, Charles Bright was able to extend these investigations until the total length under test was upwards of 2,000 miles. He was able to determine the practicability of working through a cable of the length required to connect Ireland with Newfoundland.

To avoid interrupting the traffic, the experiment had to be made during the night, or on Sundays. Hence, on many occasions, Charles Bright was unable to return home at the end of a heavy day's work.

The inductive effects observed in the earlier stage of these trials—then an entirely novel phenomenon —and the consequent retardation of the current, were described in a paper read by Mr. Edward Bright, accompanied by experiments, at the 1854 British Association Meeting held in Liverpool.[1] In 1855, the practical results of these researches were included in a patent taken out by Charles Bright and his brother for signalling through long distances of gutta-percha-insulated conductors, by the employment of alternating currents.

During these years the Magnetic Company's system had been completed by Charles Bright through Ireland, and extended to the West Coast at various points, including Limerick, Galway, Sligo, Portrush, Tralee, and Cape Clear Island. The wires were erected mostly on the railways and under exclusive agreements ; and a few miles extension from one or other of these stations would

[1] "The Retardation of Electricity through long Subterranean Wires," see *Reports of the British Association for the Advancement of Science.*

suffice to connect the system to an Atlantic cable. Whilst Charles Bright was engaged on the completion of his experiments preliminary to the great Atlantic work, his brother, accompanied by some of the "Magnetic" staff, took an opportunity of surveying, in the summer of 1855, the westernmost part of the Irish coast in a fishing smack, for the purpose of ascertaining the best landing place for the proposed cable. The main conditions required by Charles Bright were :—

(1) Freedom from anchorage.

(2) Shelter from rough weather.

(3) A smooth bottom for the heavy shore end of the cable, and the deeper part at the approach clear of rocks.

Obviously the nearer to Newfoundland the better. Various small harbours and bays between Bantry Bay and Ventry Harbour were examined; also Doulas Bay, Valentia, leading up to the Cahirciveen on the mainland. The country in the neighbourhood of Dingle Bay was in a most poverty-stricken condition, the inhabitants in many cases living, half naked, in holes scooped in the side of the hills, with the entrance partly thatched over, much as the troglodytes or "cave-dwellers" of early days; the land itself mainly consisted of barren hills and boggy tracts.

It is no light matter sailing off that coast at any time in a small vessel, however seaworthy. As luck had it, rough weather prevailed just then. Thus the company composing the expedition were often drenched to the skin. Valentia Harbour was eventually considered to best comply with the aforesaid requisites,[1] besides being almost the nearest point to the Newfoundland outstretched hand of the American continent; and Edward Bright reported accordingly to his brother Charles.

Whilst Ireland was thus telegraphically equipped as the great stepping-stone on this side of the ocean, matters on the American side were not so far advanced. The work there was much heavier, for it involved a long land telegraph across Newfoundland over a very wild country.

In 1852, Mr. Frederick Newton Gisborne, an English engineer—previously engaged in constructing the Nova Scotia telegraph lines—in concert with a small American syndicate (headed by Mr. Horace B. Tebbets, of New York) obtained an exclusive concession for thirty years in Newfoundland, with a grant of land and other privileges, subject to the erection of a line between St. John's and Cape Ray in the Gulf of St. Lawrence, whence news and messages were to be passed to and from Cape Breton, on the other side of the Gulf, by steamer or carrier pigeons. A few miles of cable were made in England, and laid between Prince Edward Island and New

---

[1] The selection has been abundantly justified, as may be gathered from the number of Atlantic cables since landed there, or in the immediate vicinity.

Brunswick with much difficulty. Mr. Gisborne then surveyed the route for the Newfoundland line, and even erected about forty miles of it. At this stage, his American associates stopped supplies.

When in New York in 1854, however, Gisborne was fortunately introduced to Cyrus West Field, a retired merchant. Mr. Field was a man of sanguine temperament, and intense business energy;[1] and having caught on to the idea of

MR. CYRUS WEST FIELD

the Atlantic cable, had the acumen to recognise the importance of confirming and turning to useful purpose the exclusive rights granted to Mr. Gisborne. He formed a strong syndicate with half a dozen friends, and proceeded with Mr. Gisborne

---

[1] In his 1887 Inaugural Address to the Society of Telegraph Engineers (now the Institution of Electrical Engineers), Sir Charles described Mr. Field as "rapid in thinking and acting, and endowed with courage and perseverance under difficulties—qualities which are rarely met with." (See *Jour. I. E. E.*, vol. xvi. p. 7).

NEWFOUNDLAND TELEGRAPH STATION, 1855

to St. John's, where they procured a concession with improved terms—the exclusive right being extended to fifty years, with a guarantee of interest on £50,000 of bonds, £5,000 in cash, towards a bridle path along the line of land telegraph—and finally a grant of fifty square miles of land, together with fifty miles more on completion of the cable across the ocean. This was followed up by the Syndicate (registered as the New York, Newfoundland and London Telegraph Company) also acquiring sole rights for landing cables on Prince Edward's Island, Cape Breton Island, Nova Scotia, and the State of Maine—some distance to the westward, between St. John's and the boundary of the State of Massachusetts, which latter refused to grant a similar concession. These collectively formed such a range of exclusive landing privileges as would debar competition for a considerable period in favour of those who might be persuaded to risk their money in what was undoubtedly, at the time, a hazardous undertaking.

Armed with this apparent monopoly, but as his brother, Mr. Henry Field, expressed it, "with no experience in the business of laying a submarine telegraph," the presiding genius of the Newfoundland Company was despatched to England at the end of 1854, where he ordered a cable of about eighty miles, to span the Gulf of St. Lawrence between Cape Ray and the Island of Cape Breton. There he became acquainted with Mr. John Watkins Brett, who, with his brother Jacob, had taken the foremost part in establishing the first lines to France and Belgium.

MR. JOHN WATKINS BRETT

In the spring of 1855, Mr. Brett took £5,000 in shares and bonds in the "Newfoundland" Company, thus becoming a partner on equal terms

CABLE-LAYING IN THE GULF OF ST. LAWRENCE, 1855

with Mr. Field and the other members of the Syndicate.

The attempt to lay the Cape Breton cable

was a failure, owing to rough weather, when the sailing vessel containing it was in tow of a steamer. But in the following year Mr. (afterwards Sir Samuel) Canning successfully laid it from a steamer ; and in 1856 the aerial land line was stretched across Newfoundland.[1]

Thus, then, the series of stepping-stones were now also completed on the American side.

SECTION II

Formation of the Company and Construction of the Cable

The next steps towards the realisation of the enterprise, in which Charles Bright's energy was centered, had better be told in his own words :[2]

"In July, 1856, Mr. Cyrus Field, the deputy chairman of the New York and Newfoundland Telegraph Company, left America for London, empowered by his associates to deal with the exclusive concession possessed by that Company for the coast of Newfoundland and other rights in Nova Scotia. He had been here before about

[1] The distance to be spanned was nearly 400 miles, through a wild, almost uninhabited country. For the most part an unbroken forest, the trees had to be cut down here to make room for the telegraph posts.

[2] Sir Charles Bright's Presidential Address, to the aforesaid Society of Telegraph Engineers and Electricians, 1887.

telegraph business, and I had discussed the Atlantic line with him in the previous year.

"On the 29th September, 1856, an agreement was entered into between Mr. Brett, Mr. Field, and myself, by which we mutually, and on equal terms, engaged to exert ourselves *with the view, and for the purpose of, forming a Company for establishing and working of electric telegraphic communication between Newfoundland and Ireland, such Company to be called the 'Atlantic Telegraph Company,' or by such other name as the parties hereto shall jointly agree upon.*"

We here reproduce the signatures as they are at the foot of this agreement:—

As the business man, Mr. Field naturally appeared before the London public, and was brought into communication with the Press authorities, more

CHARLES TILSTON BRIGHT
(*Age* 24)
When Engineer to the Atlantic Telegraph Company

than his *confreres*. Thus, it was perhaps but natural that his name should sometimes be more especially put forward, and that he should have been accorded more than his fair share of credit in connection with this great work. Not only has he been frequently spoken of as though he were the sole projector, but he has even been referred to, by the uninitiated, as the engineer.

The above "promoters and projectors" were a little later joined by Mr. Edward Orange Wildman Whitehouse, previously a medical prac-

MR. WILDMAN WHITEHOUSE

titioner at Brighton. Mr. Whitehouse had interested himself in electrical matters in a scientific way for a number of years. He was a gentleman of very high intellectual and scientific attainments, and a most ingenious and painstaking experimenter. Having indeed been engaged for some time previous upon experiments[1] similar to those, on which the

[1] Set forth in the course of B.A. Papers read at the Liverpool, Glasgow, and Cheltenham Meetings, respectively, of 1854, 1855, and 1856.

brothers Bright had been at work, with a view to overcoming the difficulties incidental to long distance ocean telegraphy. Upon Mr. White-house becoming a party to the Atlantic scheme, he also, by arrangement with Charles Bright, made a series of investigations on the Magnetic Company's lines.

About the same time, Professor Samuel Morse, LL.D., electrician to the "New York and New-

PROFESSOR S. F. B. MORSE, LL.D.

foundland" Company, came over to England to watch matters in their interest. From conclusions arrived at from experiments made at the office of the "Magnetic" Company, in Old Bond Street (on October 9th, 1856), in his presence, this distinguished American electrician expressed himself thoroughly satisfied as to the possibility of carrying the scheme to a successful issue.

The time had now come for action. As a result of considerable discussion, the two Governments concerned came to recognise the grandeur and feasibility of this undertaking for linking together the two English-speaking nations, and the benefits it would confer upon humanity. Both English and United States Governments gave a subsidy, which jointly amounted to eight per cent. on the capital, but payable only while the cable worked.

The Atlantic Telegraph Company was registered on the 20th October, 1856.

The Magnetic Company, under the management of Charles Bright, had proved a success from its foundation in 1852. The lines had been constructed, and the staff trained, under his supervision, while the improved telegraphic apparatus and appliances employed, were devised by him. The headquarters were in Liverpool, and the shareholders were composed of the leading merchants and shipowners there, as well as in Manchester, London, Glasgow, and Dublin. The Magnetic Company's Board was composed of practical business men, who fully appreciated the immense advantages which direct communication with America would bring them—not only as regards their trade, but on account of increased traffic over the "Magnetic" lines, which alone extended through Ireland. The directors had also acquired thorough confidence in their comparatively youthful chief officer, and appreciation of the value of the experiments and scientific investigations, which he had carried out.

The first meeting of the "Atlantic" Company was convened for the 12th November, 1856, at the Underwriters' Rooms in the Exchange, Liverpool, by a small circular on a half-sheet of notepaper, issued by Mr. Edward Bright from the Magnetic Company's chief office. Most of the enterprise, influence, and wealth of the town were represented, and the inspiriting addresses of Messrs. Field and Brett, accompanied by the scientific explanations (and answers to questions) of Charles Bright, were exceedingly well received. A portion, however, of Mr. Field's speech was somewhat unpractical, and produced a rather curious impression. Though well meant he said *inter alia* that :—

"It was proposed to have two vessels to lay the cable. These vessels should proceed together, each having half of the cable on board, to some point midway across the Atlantic ; and then, having carefully connected the ends of the wire together, and having telegraphed from one to the other, the vessels, having each an end on board, should sail, the one for Newfoundland, and the other for Ireland. As they expected to have the cable submerged between Monday morning and Saturday night, it should not cause them to violate the Sabbath." [1]

The idea was pious certainly, but the majority of the shipowners and underwriters present pro-

[1] *Northern Times*, Liverpool, November 13th, 1856.

bably preferred Sunday as a day of sailing : moreover, there could naturally be no question of waiting till Monday in mid-ocean, if the expedition was ready to start with suitable weather on any other day. Mr. Field, however, was scarcely known in the North, and so much enthusiasm had been aroused by the experiments and explanations already alluded to, that the directors and shareholders of the "Magnetic" Company, with their friends in Liverpool, London, Manchester and Glasgow, came forward at once, and in a few days provided the necessary capital.

The public lists were opened at the Magnetic Company's chief offices in the Exchange, Liverpool, and at their other principal offices. The first to put down their names (apart from the formal subscribers to the Memorandum of Association) were Charles Bright and two old friends, Mr. Joseph Hubback (Mayor of Liverpool) and Mr. Charles Pickering (of Messrs. Schroder & Co.), the two former for £2,000 each, and the latter for £6,000. Subsequently, Mr. J. W. Brett, who was a man of wealth, took up shares to the value of £25,000, Mr. Field following his example for a similar amount.

The formation of the Company was absolutely unique at the time, and formed a fit complement to the grandeur of the enterprise. There was no promotion money ; no prospectus was published. There were no advertisements, no brokers, and no commissions were paid; nor were there either board of directors or executive officers.

The election of a Board was left to a meeting of shareholders, to be held after the allotment of shares had been made by a provisional committee. Any remuneration of the projectors was made wholly dependent upon, and subject to, the profits of the shareholders, amounting to 10 per cent. per annum, the surplus being then divided between the promoters and the Company.

To show the interest taken in the scheme, even those entirely unconnected with business took shares, among others being the widow of Lord Byron, and Mr. Thackeray, the author.

Mr. Field had reserved £75,000 for American subscription, for which he signed, in addition to what he took for himself; but his confidence in his compatriots turned out to be greatly misplaced. The result has been thus told by Mr. Henry Field :—

"In taking so large a share it was not his intention to carry this load alone. It was too large a proportion for one man. But he took it for his countrymen. He thought one-fourth of the stock should be held in this country (the United States), and he did not doubt, from the eagerness with which three-fourths had been taken in England, that the remainder would be at once subscribed in America."

It was only, in fact, after much trouble that subscribers were obtained in America for a total of twenty-seven shares—or less than one-twelfth of the total capital. And so again the truth was proved of the ancient adage (much to Mr. Field's misfortune as it happened) that no man is a prophet in his own country.

The faith of the Americans in the project proved to be small, for (notwithstanding their confessed enthusiasm) they certainly did not rise to the occasion, and when they did so it was only after considerable pressure.

Mr. Edward Bright, however, "placed" a part of Mr. Field's holding amongst "Magnetic" shareholders and other friends, and this somewhat lightened his load.

The negotiations with Government led to important results, which were thus embodied in a a letter :—

Treasury Chambers,
November 10*th*, 1856.

SIR,—

Having laid before the Lords Commissioners of Her Majesty's Treasury your letter of the 15th ultimo, addressed to the Earl of Clarendon, requesting certain privileges and protection in regard to the line of telegraph which it is proposed to establish between Newfoundland and Ireland, I am directed by their Lordships to inform you that they are prepared to enter into a contract, based upon the following conditions, viz :—

1. It is understood that the capital required to lay down the line will be (£350,000) three hundred and fifty thousand pounds.

2. Her Majesty's Government engage to furnish the aid of ships to take what soundings may still be considered needful, or to verify those already taken, and favourably to consider any request that may be made to furnish aid by their vessels in laying down the cable.

3. The British Government, from the time of the connection of the line, and so long as it shall continue in working order, undertakes to pay at the rate of (£14,000) fourteen thousand pounds a year, being at the rate of four per cent. on the assumed capital, as a fixed remuneration for the work done on behalf of the Government, in the conveyance outward and homeward of their messages. This payment to continue until the net profits of the proposed Company are equal to a dividend of six pounds per cent. per annum, when the payment shall be reduced to (£10,000) ten thousand pounds a year, for a period of twenty-five years.

It is, however, understood that if the Government messages in any year shall, at the usual tariff charged to the public, amount to a larger sum, such additional payment shall be made as equivalent thereto.

4. That the British Government shall have a priority in the conveyance of their messages over all others, subject to the exception only of the Government of the United States, in the event of their entering into an arrangement with the Telegraph Company similar in principle to that of the British Government, in which case the messages of the two Governments shall have priority in the order in which they arrive at the stations.

5. That the tariff of charges shall be fixed with the consent of the Treasury, and shall not be increased, without such consent being obtained, as long as this contract lasts.

I am, Sir,

Your obedient servant,

JAMES WILSON.

THE ATLANTIC CABLE

The first meeting of shareholders took place on the 9th December, 1856, and a board of directors was elected as follows :—

LONDON.

George Peabody, Esq.   C. M. Lampson, Esq.
Samuel Gurney, Esq.   T. H. Brooking, Esq.
T. A. Hankey, Esq.   G. B. Carr, Esq.
                J. W. Brett, Esq.

LIVERPOOL.

W. Brown, Esq.   Edward Johnston, Esq.
G. Maxwell, Esq.   Henry Harrison, Esq.
Robert Crosbie, Esq.   C. W. H. Pickering, Esq.

MANCHESTER.

John Pender, Esq.   James Dugdale, Esq.

GLASGOW.

Sir James Anderson, M.P.[1]   W. Logie, Esq.
            Professor William Thomson.

The first chairman was Mr. Brown, M.P. (afterwards Sir William Brown, Bart.), Mr. Samuel Gurney, M.P., and after him Mr. T. H. Brooking, being deputy chairman, whilst Mr. Lampson (later Sir Curtis Lampson, Bart.) took the vice-chair. Mr. Brown was elected chairman mainly on account of extensive business relations with the United States ; but on his subsequent retirement, the chair

---

[1] Not to be confounded with Captain (afterwards Sir James) Anderson who commanded the *Great Eastern* during the Atlantic Cable Expeditions of 1865 and 1866.

was occupied by the Right Hon. James Stuart-Wortley, M.P.

To instance the large part taken by the Magnetic Company in this undertaking, no less than ten of the Board of the "Atlantic" were also directors or shareholders of the "Magnetic," prominent amongst them being Mr. Brett. Several names destined to have a remarkable influence subsequently on the development of submarine enterprise may here be referred to.

Mr. Pender, afterwards Sir John Pender, G.C.M.G., M.P., then a "Magnetic" director, afterwards took a leading part in the vast extensions that have followed to the Mediterranean, India, China, Australia, Brazil, the Cape, and a large proportion of the other existing cables, including several of the subsequent Atlantic lines. He was the first to preside over the affairs of the Telegraph Construction Company, and was actually chairman up to the time of his death of about a dozen cable companies, representing some £30,000,000 of capital, and mainly organised and conducted successfully through his influence, foresight, and indomitable energy.[1]

[1] Since his death, in the summer of 1896, a memorial has been set on foot, headed by the Marquis of Tweeddale, to his honour. Moreover, his eldest son, Mr. James Pender, M.P., has since been created a baronet—no doubt largely in recognition of his father's services to submarine telegraphy and to his country.

THE ATLANTIC CABLE

Professor William Thomson, of Glasgow, now Lord Kelvin, G.C.V.O., F.R.S., also was a tower of scientific strength on the Board. He had been from the outset an ardent believer in the Atlantic Cable ; indeed he had stated his views as to its practicability before the Royal Society in the year 1854. His acquisition as a director, was destined

PROF. WILLIAM THOMSON, F.R.S. (*now* LORD KELVIN).

to prove of vast importance in influencing the development of trans-oceanic communication ; for his subsequent experiments on the Atlantic Cable during 1857–58 led up to his invention of the mirror galvanometer and signalling instrument, whereby the most attenuated currents of electricity, which were incapable of producing visible signals on other telegraph apparatus, were so magnified in their effect by reflection, as to be readily legible. A description of this invention will be given in subsequent pages.

Charles Bright was appointed Engineer-in-chief by the Board, with Mr. Whitehouse as Electrician. Mr. Cyrus Field became the General Manager, and later, Managing Director. Mr. Brett having (as already stated) a seat on the Board. Then Mr. George Saward, hitherto the able manager of the British Telegraph Company (which had been amalgamated with the " Magnetic " Company), undertook the duties of secretary. On taking up the position of engineer to the Atlantic Company, Charles Bright appointed Mr. John Temple, previously connected with the British Telegraph Company, to be his secretary, and later on this gentleman was attached, in a technical capacity, to the engineering staff, an appointment resulting in the greatest satisfaction to all concerned.

The chart opposite (a reproduction of the original) shows the route proposed and adopted for the cable together with the line of soundings taken, as previously stated, by Lieut. Berryman and Commander Dayman.

Charles Bright recommended a cable with a much larger copper conductor than had ever been used before, weighing in fact 3½ cwt. (392 lbs.) per nautical mile, and the same weight of gutta percha for the insulator, but he found that this

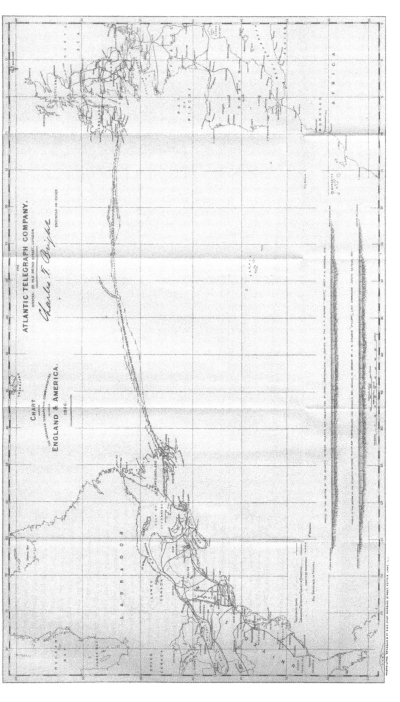

The material originally positioned here is too large for reproduction in this reissue. A PDF can be downloaded from the web address given on page iv of this book, by clicking on 'Resources Available'.

point had been settled and the contract given out before he became engineer.[1]

Indeed, a provisional committee of those registering the Company, including Mr. Brett, Mr. Field and Mr. Samuel Statham, a gentleman closely associated with the manufacture of early telegraph lines, had in their anxiety to save time, and enable the work to be carried out during the summer of 1857, entered into contracts for a cable with only 107 lbs. of copper conductor per nautical mile, and 261 lbs. of gutta percha insulation. It is true that the core specified by Charles Bright would have weighed, on the 2,500 miles of cable to be shipped, about 460 tons more; but the cable having upwards of $3\frac{1}{2}$ times the conducting power, the signalling speed he calculated on from the preceding experiments would then have been realised—not to mention that the insulation would have been more reliable. Unfortunately, those who had arranged for the smaller core were fully supported by Mr. Whitehouse's views, which moreover received entire approval from the great electrical expert—Michael Faraday,[2] as well as from Prof. Morse.

[1] On being consulted by the Government in regard to the proposed Falmouth-Gibraltar line in 1859, Bright recommended the same core as above. In this instance he had the satisfaction of seeing his recommendation adopted, though the cable was ultimately applied to connecting up Malta and Alexandria.

[2] During the discussion on a paper read by Mr. F. R.

This gentleman reported that "large coated wires used beneath the water or the earth are worse conductors—so far as velocity of transmission is concerned—than small ones; and therefore are not so well suited as small ones for the purposes of submarine transmission of telegraphic signals." Not so, however with Prof. Thomson, who had previously drawn swords with Mr. Whitehouse in connection with the latter's B.A. paper of 1854, on "Experimental Observations on an Electric Cable." Mr. Whitehouse appeared to consider a low inductive capacity as the one and only point to be aimed at in the design of a submarine conductor, without regard to the resistance offered by the wire to an electric current. In the course of a correspondence in the *Athenæum*, Prof. Thomson pointed out that the number of words which in a given time could be sent through a long submarine cable varied inversely as the *square* of the length of that cable, and not merely with the length as had been supposed by Mr. Whitehouse. Shortly afterwards, in a paper published in the *Proceedings of the Royal Society*, Prof. Thomson made public the complete theory.[1]

Window at the Institution of Civil Engineers, Prof. Faraday remarked : " The larger the wire, the more electricity was required to charge it ; and greater was the retardation of that electric impulse, which should be occupied in sending that charge forward."

[1] See the *Mathematical and Physical Papers of Sir*

THE ATLANTIC CABLE

A little later, Mr. S. A. Varley set forth in non-mathematical language, the true electrical qualifications for the working of a submarine cable. He showed in a very convincing way that conductor resistance was as much a factor in retardation as induction. He also lectured on the same subject before the Society of Arts.[1] Shortly afterwards his brother, Cromwell Fleetwood Varley, dealt with the matter in a similar spirit.[2] On his appointment as engineer, Charles Bright made every effort to get the contract altered in favour of the larger conductor which he had recommended ; but this change was not considered practicable as it would have entailed the raising of a considerable amount of further capital.

William Thomson, LL.D., D.C.L., F.R.S. (Cambridge University Press). During this discussion Mr. Whitehouse remarked that, " if Prof. Thomson's theory were correct, the dimensions of the core required to ensure the carrying capacity of the cable proving a commercial success, would be almost prohibitive and would involve a ship of correspondingly prohibitive dimensions—so much so that he doubted whether, under such circumstances, the undertaking could ever be carried out."

As is now well known, subsequent experience fully confirmed every part of Prof. Thomson's theory ; although the constants he used have been somewhat modified, and vary, of course, with the different materials used for insulation.

[1] *Proceedings of the Institution of Civil Engineers,* vol. xvii., 1858.

[2] *Proceedings of the Royal Institution,* vol. v., p. 45.

What the actual making of the cable alone entailed, the following detailed description will serve to show :—

The conductor weighing 107 lbs. per nautical mile, consisted of seven strands of copper wire, each of No. 22 gauge, covered with 261 lbs. of gutta percha, in three separate layers, 1·8 to No. ·00 B.W.G. ⅜ inch. This insulated core was then served spirally with hemp yarn saturated with a preservative composition of tar, pitch, linseed oil, and wax. The core was then protected

THE DEEP SEA CABLE

by an armour of eighteen iron strands, each composed of seven fine wires, also of No. 22 gauge, wound around [1] in a long spiral.[2] The finished

[1] The manner in which the specimens had been given out for tender by the original provisional committee to the different firms, led to the wires being eventually applied with an opposite lay at the two sheathing factories. On Charles Bright becoming engineer he learnt what had been done by the Committee. The matter was not, however, considered to be serious, neither was it found so afterwards.

[2] This particular type of iron sheathing was adopted partly at the suggestion of the late Mr. Isambard Kingdom Brunel, F.R.S., one of the greatest engineers of the

cable then received a coating of a cold mixture (referred to further on) of tar pitch, and linseed oil.[1]

Its weight in air was about 20 cwt., and in water $13\frac{1}{2}$ cwt. with a breaking strain of about $13\frac{1}{2}$ tons.

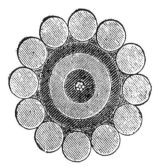

THE SHORE-END CABLE

For each end approaching the shore, the sheathing (see illustration) consisted of twelve wires of

day : Mr. Glass also strongly recommended it. Nowadays, such wires would be considered too fine, besides the stranding being undesirable ; but at that time there was great difficulty in obtaining a high class of wire from larger gauges.

[1] A cable of this pattern with not only a small surface but an extremely smooth one, was necessarily very free from friction. It would therefore sink rapidly—*i.e.* at an angle such as would ensure the line dropping into and lying evenly in the irregular surface of the sea-bottom. Moreover, the fine stranded wires provided a flexibility which rendered the line capable of adapting itself, without strain, to circumstances in a way that a more rigid cable would be absolutely unequal to ; besides being light and strong for the purposes of recovery and repair.

No. o gauge, making the total weight of over eight tons to the mile. This type was adopted for the first ten miles from the Irish coast, and for fifteen miles from the landing at Newfoundland, at both of which localities rocks had been found to abound plentifully.

Only six months was allowed to the manufacturers to complete the 2,500 miles. This involved the preparation and drawing of 17,500 miles of copper and stranding it into the 2,500 miles of conductor. Then the three separate coatings of gutta percha had to be applied round it, and subsequently the yarn. Finally 315,000 miles of charcoal-iron wire had to be drawn, laid up into 45,000 miles of strand, and the core had to be bound around with it.

The entire length of copper and iron wire employed was therefore 322,500 miles—enough to engirdle the earth thirteen times, and considerably more than enough to extend from the earth to the moon.

Various experiments with sample lengths of different iron wires made up into cable were carried out at the works of Messrs. Brown, Lenox & Co., the famous engineers, and makers of cable laying and repairing appliances.

Deaf to their engineer's entreaties,[1] the Board,

[1] Besides recommending a far larger conductor, Bright had also urged that about a year should be devoted to a

on meeting, merely confirmed the preliminary contracts with the Gutta Percha Company for the core, as also with Messrs. Glass, Elliot & Co., and Messrs. R. S. Newall & Co., the former to cover half the cable with its outer sheath at East Greenwich, and the latter to treat the other half at Birkenhead, both these firms being well-known as the most experienced manufacturers of the day.

MANUFACTURE OF THE INSULATED CORE

This subdivision of labour (by giving half the contract to Messrs. Newall) was decided upon in order, in the first place, to complete the work within the appointed time ; and secondly, with a

number of experimental investigations in connection with all the branches of science—engineering, nautical and electrical—involved by this enormous undertaking.

view to checking threatened opposition. This was a somewhat prejudicial arrangement, as it precluded any testing or trial of the entire length until the ships met at Queenstown. However, Mr. Field and some of his associates were anxious to hurry on. Their sole aim was to get the immense length of cable made and laid the fol-

SERVING THE CORE WITH HEMP-YARN

lowing summer—a few months only after it was actually ordered.

The construction of the line was commenced with all despatch at the three factories. The manufacturers of the core were closely supervised at the Gutta Percha Company's Works at Wharf Road by their manager, Mr. Statham. An idea of this operation as performed in those days may be gathered from the accompanying illustration.

Similarly, at the Greenwich Works both Mr. (afterwards Sir Richard Atwood) Glass, and Mr. (afterwards Sir George) Elliot were indefatigable in their efforts to attain perfection of construction, and to complete their order within the required time. Mr. Newall evinced a corresponding amount of zeal in the work upon his portion of the cable at Birken-

APPLYING THE IRON SHEATHING WIRES
[From *The Illustrated London News*]

head. The view here shows the sheathing operation—*i.e.* the application of the stranded iron wire—as carried out during the construction of this cable.

The *Illustrated London News* published a number of views of the manufacture of this, the pioneer, Atlantic cable, as well as general descriptions of the machinery employed, and some of these sketches are reproduced from their pages by kind permission of the proprietors.

Finally, we have here an illustration of the cable being coiled down, as fast as made, into the large iron tanks in the factory, ready for shipment.

When once the wheels had been fairly set in motion, it was necessary for Charles Bright to gather round him a competent staff of engineers,

MR. (NOW SIR SAMUEL) CANNING

ready for the expedition. First of all, as his chief assistant, he secured the services of Mr. Samuel Canning, who had laid the Gulf of St. Lawrence Cable, for Messrs. Glass & Elliot, in the preceding year. The next place was filled by Mr. William Henry Woodhouse, who had laid cables for Mr. Brett in the Mediterranean. Then came Mr. F. C. Webb, who had probably been as-

134

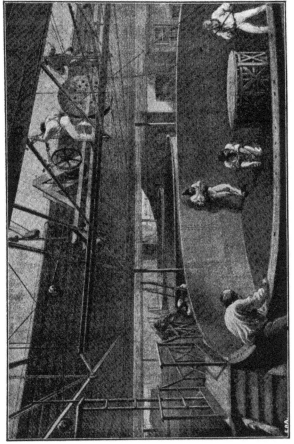

COILING THE FINISHED CABLE INTO THE FACTORY TANKS

sociated—in one capacity or another—with more early cable work than any other single telegraph engineer; he took an appointment for special duties on Charles Bright's staff.[1]

Finally, Mr. Clifford joined. He was a cousin of the Taylors,[2] and was in this way introduced to the undertaking, besides being a mechanical en-

MR. HENRY CLIFFORD

gineer of considerable experience. It should also be mentioned that Mr. John Kell—a captain in the merchant service, who had assisted in the laying of cables in the Mediterranean—was attached to

[1] In his early days, Mr. Webb was engaged on marine surveys in H.M.S. *Porcupine* as well as on railway work.

[2] Though only related to his wife's family (and not to Bright himself), strangely enough a strong likeness was often seen between Mr. Clifford and the subject of our memoir.

the staff, his duties being to look after the men at their work, and to prepare accommodation on board ship.[1]

A few extracts from Charles Bright's diary may here be of interest, as showing the arduous and constant vigilance necessary in superintending the manufacture.

1st Jan., 1857.—At Greenwich (Glass & Elliot's), saw sample cable 60 ft. long spun off. Considered about keeping the wire in tank either always covered with water or always dry. Appointment with Edgington's *re* tarpauling for covering coils. Talked with Canning as to undertaking part of charge of paying out machinery. Appointment for test cable.

Saturday, 3rd.—To Brown & Lenox's, at Millwall, at one, to test cable with Glass. Two samples broke off at the clamp ; not fair trial ; fresh appointment for Tuesday.

Tuesday, 6th.—To Brown & Lenox's to test cable ; stood up to 3 tons 11 cwt. Then to Greenwich, testing joints.

Monday, 19th.—To Gutta Percha Works in morning. then to Greenwich. Spinning started with one machine. Discussion as to tarpaulin covering. Edgington's want £350 for six months' rent of tarpaulins.

Friday 23rd.—Tar-pitch mixture (cold) answers very well for coating (with a brush) outside of cable, as a preservative against rust.

[1] There were a number of assistants on the engineering staff, but the only others of which any record can be found were Mr. Suter, Mr. Windle, and Mr. O. Smith, all of whom had been previously engaged on cable work.

THE ATLANTIC CABLE

Jan. 27th.— 3 barrels tar,

½ barrel pitch,

12 lbs. beeswax,

6 gallons linseed oil.

} Preservative mixture decided on.

Twelve or thirteen gallons per mile.

All the contractors concerned in this work were ready with their supply within the time stipulated.

Among the illustrious visitors at Greenwich during the construction were the Prince of Wales, and Prince Alfred, now Duke of Edinburgh; both evinced a lively interest in the work, and carefully studied each stage of the manufacture, young Bright having the honour of acting as "showman."

In January, 1857, the celebrated poet, Martin Tupper, contributed the following inspiring verses in praise of the undertaking :—

> World! what a wonder is this
> Grandly and simply sublime,
> All the Atlantic abyss
> Leapt up in a nothing of time!
> Even the steeds of the sun,
> Half a day panting behind,
> In the flat race that is run,
> Won by a flash of the mind.
>
> Lo! on this sensitive link—
> It is one link, not a chain—
> Man to his brother can think,
> Spurning the breadth of the main ;
> Man to his brother can speak,
> Swift as the bolt from a cloud,
> And where its thunders were weak,
> There his least whisper is loud !

Yea, for as Providence wills,
  Now doth intelligent man
Conquer material ills,
  Wrestling them down as he can,
And by one weak little coil,
  Under the width of the waves,
Distance and time are his spoil
  Fetter'd as Caliban's slaves !

Ariel ?—right through the sea,
  We can fly swift as in air ;
Puck ?—forty minutes shall be
  Sloth to the bow that we bear ;
Here is Earth's girdle, indeed,
  Just a thought-circlet of fire,
Delicate Ariel freed
  Sings, as she flies, on a wire !

Courage, O servants of light !
  For you are safe to succeed ;
Lo, ye are helping the right,
  And shall be blest in your deed :
Lo, ye shall bind in one hand,
  Joining the Nations as one,
Brethren of every land
  Blessing them under the sun !

This is Earth's pulse of high health
  Thrilling with vigour and heat,
Brotherhood, wisdom and wealth,
  Throbbing in every beat ;
But you must watch in good sooth,
  Lest to false fever it swerve,
Touch it with tenderest truth
  As the world's exquisite nerve !

THE ATLANTIC CABLE

Let the first message across,
  High-hearted Commerce, give heed,
Not be of profit or loss,
  But one electric indeed ;
Praise to the Giver be given
  For that He giveth man skill,
Praise to the great God of Heaven,
  Peace upon earth and goodwill !

Section III

Ships, Stowage, and Departure for Valentia

Charles Bright was but twenty-four years old at
the time when he was appointed chief engineer to

MR. C. T. BRIGHT
[From the *Illustrated London News* at this period]

carry out perhaps the most important and certainly
the most surprising enterprise of the century. His
task was to enable the people of the two great

Continents to speak together, in a few moments of time, though separated by a vast ocean!

The work and responsibility devolving upon him were enormous, and probably few young engineers have at his age been placed in a position of such heavy responsibility.

Improved paying out machinery to suit the great depths required to be devised.

He had to select ships suitable to carry two thousand five hundred miles of cable, and to prepare them to receive it, together with the requisite machinery, and so arrange the distribution of weight as to keep them fairly in trim.

In addition there was the necessity for almost constant attendance on the directors, sitting as they did almost daily, and the preparation of frequent reports. Another task was to tone down the excited feelings of endless would-be inventors.

It was just about this time—the end of 1856—that the scene of young Bright's home was changed from Southport, near Liverpool, to the Cedars,[1] near Harrow.

Soon after becoming engineer to the under-

---

[1] Here the family remained till the beginning of 1861. The view given opposite—the only one available—shows the house as it now stands, greatly improved by its present owner, Mr. T. F. Blackwell, J.P., an old friend of the Brights.

THE CEDARS, HARROW WEALD

taking, in conjunction with the authorities of the Admiralty, he had visited and inspected various ships. Eventually H.M.S. *Agamemnon* (Master-in-Command C. T. A. Noddall, R.N.), was selected and placed by Government at the service of the Company. She proved to be admirably adapted, by her very peculiar construction, for the service of receiving the cable. Her engines were quite near the stern, while amidships she had a magni-

COILING THE CABLE ABOARD

ficent hold, forty-five feet square, and about twenty feet deep. In this capacious receptacle nearly half the cable was stowed away, an operation which Mr. Woodhouse superintended on behalf of his "chief." An idea of this may be gleaned from the accompanying sketch,[1] and a general view of her

[1] With characteristic forethought, Charles Bright devised a machine for automatically coiling (or uncoiling)

lying off the works taking in cable is shown in the full-page illustration. She was a screw propelled line-of-battle ship of ninety-one guns and one of the finest in our navy. She had seen service, too, having figured in history as the flagship of Admiral Lyons, when he took her up to within a few hundred yards of Sebastopol, at the bombardment, a couple of years before. She was now to take part in an equally, if not more glorious, achievement. She was, indeed, to do more during her coming mission to bring about the reign of peace—by drawing together in closer communion the several nations of the earth—than any man of war was ever called to do, before or after.

On their own side the American Government, after five months' hesitation, sent over the largest and finest ship of their navy, the U.S. frigate *Niagara*, a screw-corvette, which with her tonnage of 5,200, exceeded in size our largest line-of-battle ship. She was a lovely model—as may be seen from our illustration—and sat on the water like a yacht, although as much as 375 feet long and 56 feet beam—utterly unlike the then recognised ideas

the cable when short of men. This ingenious contrivance is fully described in the Patent Specification No. 990, A.D. 1857, bearing his name. As it turned out, however, there was no need for its use, though nowadays, with the constant fear of strikes and labour troubles before us, it would seem to be a useful machine to have at hand in case of emergencies.

H.M.S. *AGAMEMNON* TAKING THE CABLE ABOARD

of a great vessel of war.[1] She had been constructed by George Steers, and somewhat on the lines of the famous yacht *America*, which he had then recently built. The latter had beaten all our best at Cowes, and had carried off that challenge cup to the States which Lord Dunraven and so many others have pluckily, but so far in vain, endeavoured to wrest back again for our own.

The *Niagara* was commanded by Commodore W. L. Hudson, one of the most experienced officers of the U.S. Navy. Mr. Field and Professor Morse took passage in her to England.

Although her decks had been partly cleared at Brooklyn of bulkheads and other impediments to cable stowing, Charles Bright found there was not room enough, and had to explain to the Commodore that some further sweeping alterations would be required. In reply, Captain Hudson strongly urged that there was space already if the cable was stowed in long "flakes" along the deck, instead of in circular coils. He was not persuaded to the contrary until young Bright arranged with him to flake some away in his own fashion. It was drawn out by a "donkey-engine," when it repeatedly kinked; and Captain Hudson had to give up his point, consenting, with much heart-

---

[1] The official "fighting" description of her at the time was as follows :—" Carries twelve Dahlgren guns, weighing 14 tons each, 11 inches in the bore, throwing a solid shot of 270 lbs., or a shell of 180 lbs., a distance of four miles."

tearing, to have his beautiful ship cut still more to pieces, and, as he expressed it, "scooped out like a crab." A considerable part even of the officers' wardroom had to be appropriated. These transformations, what with the cutting and boring of decks, and then cumbering them with machinery —great buoys and other cable gear—certainly effected a woeful change.

As a consort, the U.S. paddle frigate *Susquehanna* was also detailed for the expedition, commanded by Capt. J. R. Sands, U.S.N. H.M.S. *Leopard* (Capt. J. F. B. Wainwright, R.N.), was similarly provided by our Government; whilst H.M. sounding vessel *Cyclops* (Capt. Joseph Dayman) was to precede this little fleet—nicknamed the "Wire Squadron"—to show the way.

During the short time left, Charles Bright devised apparatus for paying out the cable on a somewhat different principle to that, which had hitherto been in use for laying cables in comparatively shallow water. This was rendered necessary on account of the different conditions. Moreover, the apparatus hitherto in vogue was of a rather primitive kind, consisting of a drum, round which the cable was coiled several times, with a brake strap surrounding it, regulated by a hand lever upon a somewhat "rule of thumb" system. This arrangement had repeatedly broken down, notably in 1854 in the Mediterranean, when the cable slipped upon the surface of the break-drum used to check it, and flew out of the vessel with great force, cutting its way through the bulwarks of the ship in its passage. The same trouble of the cable surging and "taking charge," with the above rough and ready appliances, also occurred between Sardinia and Algeria in the following year.

Bright's machinery for regulating the egress of the cable from the laying vessels, was constructed with a view to (1) the great depth of water to be passed over, (2) the constant strain, and (3)

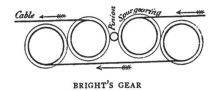

BRIGHT'S GEAR

the number of days during which the operation must unceasingly be in progress.

The cable was passed over and under a series of grooved sheaves, having the bearings of their axles fixed to a framework composed of cast-iron girders bolted down to the ship's beams. The principle of it is shown in the accompanying sketch.

The sheaves were geared to each other, and to a pinion fixed to a central shaft revolving at a rate three times faster than that of the sheaves. Two friction drums upon this shaft regulated the speed of paying out, and the grooves of the

sheaves—the latter being fixed to their axles outside the framework and bearings—were so shaped
that they acted as a complete and equal support all
round for the semi-circumference of the cable under
a strain, thus also gripping it firmly, without any
pressure, however, sufficient to injure it.[1]

The degree of retardation was regulated by a
hand-wheel, the strain on the cable being indi-

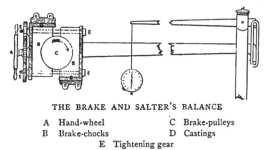

THE BRAKE AND SALTER'S BALANCE

A   Hand-wheel   C   Brake-pulleys
B   Brake-chocks   D   Castings
            E   Tightening gear

cated on a dial actuated by a form of Salter's
balance, in connection with the friction brake.

For the details and construction of this machine
(see illustration) Mr. Charles De Bergue, a well-
known mechanical engineer, was largely responsible.[2]

[1] Though theoretically correct—for maintaining the
form of the cable and for adding to the "hold" on it by
extra friction—the practical difficulties involved here had
to be discovered later.

[2] See Bright and De Bergue's Patent Specification
for "Improvements in apparatus for Laying Submarine
Telegraph Cables," No. 1,294, A.D. 1857. See also Bright's
Specification No. 990 of same year, referred to in the
Inventions Chapter forming Appendix 1, vol. ii. of this

THE ATLANTIC CABLE

One important advantage in this gear over its predecessors consisted in its having the sheaves *outside* the shaft bearings, and only half embraced by the cable, so that it could be readily cleared, in case by mischance it kinked in the hold, and came up in a tangle, as had been the case occasionally in previous undertakings. The actual manufacture of the paying out machinery, arranged by Charles Bright, was carried out by Messrs. C. De Bergue & Co., of Manchester, assisted on behalf of the Company by Mr. Henry Clifford, whose experience and talent in mechanical engineer-

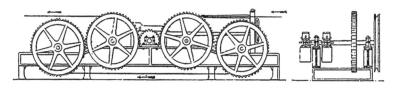

THE PAYING-OUT MACHINE OF 1857

ing here again proved of great service. A special " Picking-up " machine on similar principles was also designed by Mr. Bright and Mr. Clifford, consisting of two drums grooved to take five turns of the cable, half round each, and arranged to gear with a powerful steam engine in case of misadventure.[1] Placed slightly further forward than the paying-out gear, this apparatus was constructed without any

biography. In the above patent occurs the first mention of a cable dynamometer.

[1] This was eventually used as a paying-out apparatus in the 1858 expedition.

definite idea of recovering the cable when once laid at the bottom. It was rather intended to draw the cable back in the case of a fault during paying out, after taking over the line from the paying-out apparatus alongside.

There were also arrangements by means of which picking up could be effected from the bows and taken aft to the "winding-in" machine.

As a provision for stormy weather, Charles Bright provided two large reels on the deck, each carrying a very strong iron wire cable to attach to the telegraph line and to elongated buoys, so as to hang on to a buoyed line (with plenty of slack) in case of emergency, till paying out operations could be safely proceeded with.

In connection with this undertaking Charles Bright also invented a patent log, a wheel of which was "arranged to make and break an electric circuit at every revolution. A gutta-percha-covered wire was run up from the revolving wheel on to the deck of the ship, so that it should carry the current whenever the circuit was completed, and record there (upon a piece of apparatus provided for the purpose) the speed of the vessel."[1] Bright also placed an external guard

---

[1] This ingenious contrivance was fully described in *The Engineer*, vol. iv., p. 38, together with a general account of the proposed arrangements for laying the cable, etc.

A somewhat similar device was patented several years later by Messrs. Siemens.

over the screws of the vessels to prevent the cable from fouling with them in case any necessity should arise for backing the vessels.[1] This cage was nicknamed a "crinoline" (then in fashion), which it somewhat resembled.[2]

The construction of the cable having been divided between two competing manufacturers, who held aloof from one another, it was found that they had covered their respective portions with the outer iron wire strands in a different lay— one left-handed and the other right-handed. Both Messrs. Newall and Messrs. Glass & Elliot (formerly W. Küper & Co.) had been ordinary wire-rope makers, in which a right-handed lay was the most usual. Messrs. Glass & Elliot, however, conceived that the cable would require to be coiled down in the tank in a certain direction, in order that the turns put in during coiling should come out again as the cable emerged from the eye of the tank. Bearing in mind that a man could not well run round with the line in the left-hand direction, they arranged the lay of their cable accordingly, and made it left-handed.[3] They did not however

[1] Descriptive History published by the Directors, 1857, p. 54.—*Ibid.*, p. 55.

[2] The screw-guard may be seen at the stern of the *Niagara* in the sectional view as well as in other illustrations of her and of the *Agamemnon* (paying-out cable) further on.

[3] This has ever since been universally adopted, for the above reason.

communicate their course of action, and Messrs. Newall adhered to the ordinary rope lay.

The technical details of the above matter have been gone into at length by one of the authors of this volume more recently.[1]

Charles Bright foresaw that it would be necessary, accordingly, to make special arrangements for con-

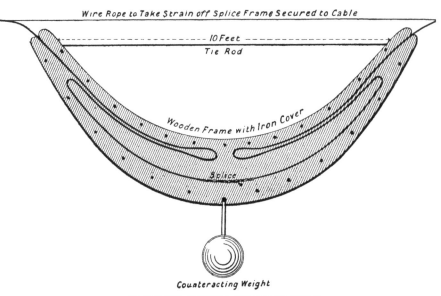

FRAME FOR HOLDING MID-OCEAN SPLICE

necting the ends of the two lengths on each ship, in order to neutralise the untwisting effect that would otherwise tend to result from the contractors having adopted opposite lays. The two ends were placed in grooves cut in a semicircular frame of wood, each with a loose projecting length. Over

[1] See article in *Engineering*, June 12th, 1894.

the grooved frame was then screwed a heavy plate of iron corresponding therewith. The short face ends being thus released from all possible strain, were jointed and spliced together outside (by what is known as a " ball-splice ") after providing a spare turn. The frame, or link-board, was weighted below, as shown in the accompanying sketch, to prevent it being turned over under tension.

It had previously been intended to start laying the cable by both ships simultaneously from mid-ocean, and Charles Bright, backed by his immediate staff—as well as by all the nautical authorities concerned—strongly urged this course, as reducing the time of laying (and, therefore, the chances of foul weather) by one half, besides having the advantage of starting with the deepest water.

Charles Bright's letter to the directors on the subject ran as follows :—

London, *July* 10*th*, 1857.
To the Directors of the Atlantic Telegraph Company.

GENTLEMEN,—I beg to lay before you my views in regard to the proposition for altering the original plan of laying the Atlantic Telegraph Cable, by starting from the Irish Coast instead of commencing from the centre as previously determined upon.

It is urged that from the possibility of a westerly gale arising during the progress of the *Agamemnon* from the centre to the eastern terminus of the cable at Valentia, there is some risk of the lowest speed at which, under such circumstances, she could make being so great as to endanger the successful laying of the line; and that it might be difficult, if not impossible, in unfavourable weather to bring her with safety so near to the coast of Ireland as to land the shore end of the cable at Valentia.

It is further advanced that the adoption of the new plan would be more desirable in the electrical department, on account of the difficulty which would exist under the original plan in ascertaining the nature of the fault in the event of any interruption of communication between the two vessels.

I have considered the former objections (which may be called the nautical difficulties) before, and have discussed them carefully with the Engineers who will assist me in laying the cable; but I have not been able so maturely to weigh the importance of the electrical points as compared with the danger which I and my colleagues see in other respects, in a change at this late period from the original plan, having only heard them suggested for the first time at the conference on Wednesday.

I think, from what passed at the meeting, that the nautical difficulty may be overcome, at all events I consider that the risk encountered by laying the cable from the Irish Coast, instead of commencing midway between Ireland and Newfoundland, is of infinitely greater importance.

By reversing the propeller of the *Agamemnon*, and by using some drag—such as that suggested by Sir Baldwin Walker—or by employing the *Leopard* in the manner proposed by Captain Wainwright; or by heaving-to the ship and allowing her to drift, the rate of the *Agamemnon* would be retarded sufficiently to deliver the cable into the sea with safety; and in this we should be assisted by having the largest and most complete coil which has ever yet been made for such a purpose.

The difficulty of making Valentia, should the weather

THE ATLANTIC CABLE

on the Irish Coast be unfavourable, is not of great consequence; for should it be necessary to part with the cable on approaching land, the comparative shallowness of the water would allow of its being easily buoyed, and connected with the shore at a more suitable time. Submarine cables are continually buoyed in the North Sea, at distances varying from ten to fifty miles from land, where they are frequently left for weeks, and are then taken up and joined to the shore.

On the other hand by starting from Ireland, we encounter the probability of losing the rope in the attempt to join it in the middle of the Atlantic, with a strain of two thousand fathoms upon it; for it must be remembered that we have been compelled, in order to keep down the bulk and weight of the cable, to make its proportions and materials such, that we cannot rely upon its sustaining a greater strain than would arise from supporting a weight of three tons and a half.

When we have to change from one ship to another, the egress of the cable must be stopped for a time, and even in fine weather this would be an operation of considerable risk ; and should there be the least wind or current, the drifting of the ships would bear upon the cable in a manner that would, to say the least, be very dangerous.

But it should be borne in mind that the occurrence of bad weather at the time when the splice must be made, would make the loss of the line an absolute certainty.

Lieutenant Maury, whose data as to the frequency of gales in the North Atlantic have been generally accepted, gives the average of gales in the month of August, between 35° and 30° long. (the part where the splice must be made), as 2, and between 30° and 25° as 1 : the other parts of our route, except those next to the shores of Ireland and Newfoundland, are estimated by comparison as 0.

Thus the 5° of longitude in the central portion of our line, where fine weather is of the most vital consequence, present almost the greatest liability to gales of the entire route. In addition to this we are informed that the probability of fogs occurring in the five degrees immediately preceding the above (that is to say, between 30° and 25° long.) may be estimated at 6, which, as will be seen from the scale below, is greater than at any other point except on approaching the coast of Newfoundland.

	55	50	45	40	35	30	25	20	15	10	
Newfoundland											. . . Ireland
Gales . .	1	1	0	0	2	1	0	0	4		
Fogs . .	6	8	8	2	0	6	1	0	3		

So that just before we enter the region where the splice must be made we run considerable risk of parting company, which would involve the danger of holding on by the cable until the vessels could meet again.

By commencing from Ireland we should double the time occupied in laying the cable, and proportionately increase the risk of losing it by meeting with bad weather.

It has been put forward as an argument in favour of the new method, that half the cable would be saved in the event of any accident arising in effecting the junction ; but under the original plan we meet with our greatest difficulty at the outset by starting midway in the deepest water. If we lay the first 100 miles—50 from each ship —safely, we have every reason to rely upon the expedition being crowned with success. If the cable is lost it will probably be in the deep water, and nearly the *whole* of the line will then be brought home.

Under the original plan we are sure of the junction being well made : we proceed to the centre, and, should not the weather be propitious, we can wait as long as may be necessary, and if any failure should occur in making the first joint, we can repeat the operation again again and again, with the loss of few miles only of the cable.

In the new plan the failure of passing the bight overboard in changing from the *Agamemnon* to the *Niagara* (which would be attended under the most favourable circumstances with very great risk), would at once lose 1,200 miles of cable; and any plan of buoying the cable to relieve the strain, such as proposed at the conference yesterday, would add to the complication of the work, which, it must be remembered, might have to be attempted at night or in bad weather.

The mode of effecting the splice in itself presents a serious difficulty in the new plan : the ends of the cable being different in the direction of their lay, each would have a tendency to unwind, and the gutta-percha-covered copper wire would then be twisted and broken off.

This must be met by serving wires reverse to the lay of each rope for at least two miles, to form a neutral length at the splice.

This would nearly double the weight of the cable in the neighbourhood of the junction, bringing a tension when the ordinary cable again passes from the vessel, more than it is capable of bearing.

For the above reasons I am, from an engineering point of view, most strongly in favour of adhering to the plan originally adopted for commencing in mid-ocean, and in this opinion I am supported by the engineers who are co-operating with me.

With regard to the electrical reasons which have now been suggested for altering our previous arrangements, I must say, that I have been completely taken by surprise at this new source of objection.

Of course, if the communication is suddenly stopped, although it is easy for the Electrician to tell whether the fault is near to his ship, or remote from it—and therefore, probably occurring within two or three miles from the other ship—still he will be unable to know what steps are being taken to remedy the defect.

This difficulty must have existed from the first, and it is peculiarly embarrassing to me that the discussion of its importance—as compared with the other conditions necessary to success—should have been deferred until a fortnight previous to our departure, when we are under so great a pressure for time, if the undertaking is to be carried out this year.

Should any defect arise between the shore and the vessel, in the new plan, the same electrical difficulty would arise, if the paying-out ship—not knowing the nature of the fault, but supposing it to be on the shore—should proceed on her voyage, continuing to pay out the cable. It is true that the chances of such an emergency do not appear so great; but we are to a certain extent trying an experiment, and although we believe, from the soundings taken from 30 to 50 miles apart, that the bottom is soft and level, there may yet be sharp rocky points where the cable may be destroyed.

In conclusion, the points which have to be decided by you, appear to me to resolve themselves into the choice between the almost certain loss of 1,200 miles of the cable by commencing from the coast of Ireland; or, by keeping to the original plan, the possibility of paying out some additional length of cable at one end, should it happen that the other end is altogether lost.

I am, Gentlemen, yours faithfully,

CHARLES T. BRIGHT,

Engineer.

THE ATLANTIC CABLE

The electrician, Mr. E. O. Wildman White-house, however—whose health did not permit him to sail with the expedition—together with the other electricians, urged that one ship commencing to pay out from Ireland, the other should continue the work when the first had used up all her cable. This course necessarily doubled the time taken in laying, and left the junction between the two cable ends to be effected in the deepest water, when it might be impracticable through rough weather. Yet such was the anxiety of the Board to keep in touch with the expedition, for daily reports of progress, that they adopted the counsel of the electricians, and decided in favour of their proposal.[1]

By the third week in July (within the course of as many weeks) the great ships had received all their precious cargo—the *Agamemnon* (Commander Noddall, R.N.) in the Thames, and the *Niagara* (Captain Hudson, U.S.N.) in the Mersey.

The final disposition of the cable on the latter vessel may be seen on next page.

Then came some farewell feastings. It seemed to be considered a suitable occasion for giving

[1] Charles Bright's plan was, however, adopted in the expedition of the following year.

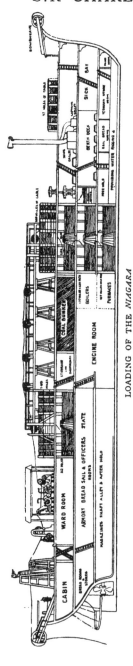

LOADING OF THE *NIAGARA*

a banquet in honour of those gentlemen about to take an active part in the laying of the cable. In the first place, the Liverpool Chamber of Commerce entertained all the officers taking passage in the *Niagara*, as well as a large number of directors of the Atlantic Company. There were also other distinguished guests, including a number of prominent local merchants.

Mr. William Rathbone, M.P., presided. In the course of his remarks the chairman said :—

Commerce would get on poorly without the aid of science and he was sure on that self-interested account, if on no other, that they would join him in cordially drinking the health of Mr. Charles Bright, the engineer of this great undertaking.

Mr. C. W. H. PICKERING (one of the directors of the Atlantic Telegraph Company), wished to express a belief, before Mr. Bright rose to respond, that if it had not been for the research, skill, and energy originally displayed by Mr. Bright when the soundings were taken, the Atlantic Cable would never have been even made. Mr. Bright was not only one of the first engineers, but he believed he was the first electrician of the day.

MR. CHARLES T. BRIGHT expressed the obligations he felt for the kind manner in which his name had been mentioned. As an engineer he had special reasons for holding in kind and grateful remembrance the merchants of Liverpool for the support they had given on various occasions to enterprises of great moment. Had it not been for their excellent chairman, and such as he, we should not have had a railway in England at the present time. At a period when to support the theory of railway trains drawn by locomotive steam engines was considered to be tantamount to the possession of but a thin partition separating one from madness (laughter), the great idea was taken up by Liverpool merchants, including some of the gentlemen then present, who had the same confidence in the statements of their engineer, Mr. Stephenson, as had now been expressed by Mr. Maxwell (cheers). He had to thank them for the very substantial support that the promoters of the scheme had received in Liverpool and in Lancashire. He believed that, if it had not been for the support received in the northern part of the county they would have had great difficulty in bringing the enterprise to a practical issue this year. It would, however (he continued), become his part in the play as an actor, and not as the speaker of a prologue, to say very little of what had been done so far, but to wait until the curtain had been raised. It might, however, be of interest that he should report progress by stating that 800 miles of the cable were now safely coiled on board the *Agamemnon* at Greenwich, and

600 miles on board the *Niagara* at Liverpool (cheers). The manufacture of the entire 2,500 miles was now completed, and he could see no reason why they should not meet together at Cork before the end of the month, and look forward to the commencement of active operations, say in August, the very best period for the undertaking (loud and prolonged cheering).

After others had spoken, the party broke up.

But a few days later, the last coil of cable having been shipped on the *Agamemnon* from the Greenwich Works, the occasion was duly honoured by a scene as unique as it was beautiful.

To quote *The Times* of July 24th :—

All the details connected with the manufacture and stowage of the cable are now completed, and the conclusion of the arduous labour was celebrated yesterday with high festivity and rejoicing. All the artisans who have been engaged on the great work, with their wives and families, a large party of the officers, with the sailors from the *Agamemnon*, and a number of distinguished scientific visitors were entertained upon this occasion at a kind of *fête champêtre* at Belvedere House, the seat of Sir Culling Eardley, near Erith. Although in no way personally interested in the project, the honourable baronet has all along evinced the liveliest sympathy with the undertaking. The manufacturers, fired with generous emulation, erected spacious tents on the lawn and provided a magnificent banquet for the guests, and a substantial one for the sailors of the *Agamemnon* and the artificers who had been employed in the construction of the cable. By an admirable arrangement, the guests were accommodated at a vast semi-circular table which ran round the whole pavilion, while the sailors and workmen sat at right angles with the chord, so that the general effect was that all lunched together, while at the same time sufficient distinction was preserved to satisfy the most fastidious. The three centre tables were occupied by the crew of the *Agamemnon*, a fine active body of men, who paid the greatest attention to the speeches, and drank all the toasts with an admirable punctuality—at least, so long as their three pints of beer per man lasted. But we regret to add that with the heat of the day and the enthusiasm of Jack in the cause of science, the mugs were all empty long before the chairman's list of toasts had been gone through. Next in interest to the sailors were the workmen and their wives and babies, all being permitted to assist. The latter it is true sometimes squalled at an affecting peroration, but that rather improved the effect than otherwise, and the presence of their little ones only marked the genuine good feeling of the employers, who had thus invited not only their workmen but their workmen's families to the feast. It was a momentary return to the old patriarchal times, and every one present seemed delighted with the experiment.

Besides Sir Culling Eardley, Bart., there were present Sir F. Thesiger, the Right Hon. Edward Cardwell, M.P., Mr. J. L. Ricardo, M.P., Professor Wheatstone, Mr. Charles Bright, Mr. Wildman Whitehouse, Mr. R. A. Glass, Commodore Noddall, and numerous other promoters, or friends, of the undertaking.

A letter was read from President Buchanan, saying that he should feel honoured if the first message should be one from Queen Victoria to himself, and that he " would endeavour to answer it in a spirit and manner becoming a great occasion."

In replying for the House of Commons, Mr. Cardwell referred to the importance of the great undertaking, and said that twenty years ago a steam voyage across the Atlantic was the great wonder of the day, but that now a marvellous agent was about to carry messages across the Atlantic with the velocity of lightning. This was a great triumph of science over difficulties. It was also a great triumph of commerce, for it was the commerce of England and America that had created the undertaking and which would successfully sustain it. But above all it was a triumph of international policy in bringing two of the greatest nations of the world into close touch with one another. Two magnificent vessels—models of the two greatest warlike fleets in the world—were about to leave our shores, not on a warlike enterprise, but to render it all but impossible that we should ever again be at war with America. The two nations had a common origin, a common language, their laws were drawn from the same fountain ; and he was sure the day would come when the present celebration would be looked upon as commencing a new era, and that every artisan and every sailor who had been engaged in the work would exult in the fact that he had assisted in completing a great bond of brotherhood.

There are some persons who venture to predict that the communication by means of the submarine tele-graph between England and America is not so near its accomplishment as the promoters and friends of the undertaking would wish ; and a number of objections have been urged, some of which completely answer each other, against the make of the cable, and the mode to be adopted in its submergence. There are some who say that the cable is too large, and others that it is too small. Some say that the inner core of the rope will become so attenuated in the process of paying it down, that it will either not convey the electrical current, or it will break altogether. Then, again, the machinery for laying it is said by some to be too strong, and that it will break, retard, or strain, the cable in the process. There never yet was a great undertaking which had not raised against it the voices of many croakers, and the pens of many critics. Of all predictions, those which prophesy failure are the safest and easiest to make; for if a mishap should occur, the prophets may take credit for their prescience, and if success should crown the undertaking, then the prediction can be claimed as merely caution and prudence in disguise. Now the persons to whom is entrusted the carrying out of this great enterprise would be guilty of the most culpable neglect if they did not avail themselves of every fact which experience might teach, or if they neglected to test the value of any theory or opinion which would be submitted to them. Who are the parties under the circumstances whom it might be desirable to associate in this undertaking? Clearly those who have had experience in the working and conducting of telegraphs. There is not a name of eminence or a person of ability in the science of telegraphy who has not in point of fact been consulted or engaged by the Atlantic Telegraph Company. There are, in the first instance, Messrs. Newall & Co., and Messrs. Glass, Elliot & Co., who have, between them, manufactured every submarine cable which has hitherto been laid down. Professor Morse has for years past devoted the whole of his ability and energy to the carrying out of the telegraph system in the United States.

SIR CHARLES TILSTON BRIGHT

Mr. Bright, our engineer-in-chief of the entire under-taking, is a gentleman whose exertions connected with electro-magnetism are well known, and deservedly appreciated by all who have paid the slightest attention to this interesting science. Did he not also successfully lay the first cable which united Ireland with Great Britain ? Mr. Brett, not less favourably known, was perhaps mainly responsible for the first line between England and France. Again, Mr. Canning has laid the cable across the Gulf of St. Lawrence, and in the Mediterranean. These gentlemen, together with Professor Thomson, Mr. Whitehouse, and several other electricians, are already engaged in connection with this Atlantic Telegraph. It is by no means impossible that some unforeseen contingency may not be provided for, but knowing in whose hands the conduct of the undertaking is placed, we think it improbable that they should not succeed. Every one connected with the enterprise appears confident of success, and the promoters are not to be daunted by imaginary difficulties. "You doubt whether we shall succeed, do you ? " said Commander Noddall to a *savant* who was predicting failure, and shaking his head ominously at some of the mechanical arrangements ; "then, Sir, on board the *Agamemnon* is no place for you." I confess to having great faith in the abilities of such practical men. From the earliest inception of this undertaking it has been the work of men of energy and of enterprise. Five years ago it was suggested by Mr. Bright, the founder of the British and Irish Magnetic Telegraph Company. He was laughed at, and the most charitable term employed towards him was that he was "a very sanguine young man." At length, however, the plan began to lose something of its visionary character, the investigation of Lieutenant Maury into the physical geography of the sea succeeded in removing the engineering difficulties which appeared to be in the way of the completion of the telegraph by proving the absence of currents, and the existence of a level bottom at the Atlantic, with a stratum likely to remain undisturbed and favourable for the reception of the cable. The "telegraphic plateau," lying between Ireland and Newfoundland, is one which "is not too deep for the cable to sink down and rest upon, and at the same time not so shallow that currents, or icebergs, or any abrading force, can derange the wire after it is once lodged upon it." This statement of Lieutenant Maury has since been verified by the soundings made by officers of British and United States navies. Having discovered a suitable plateau in which to sink the cable, it was necessary to ascertain the practicability of sending the electric current through a distance as great as that across the Atlantic. Dr. Whitehouse and Mr. Bright decided this question independently; and I understand that signals have been successfully transmitted through 2,000 miles of telegraphic wire at the rate of 270 per minute, or about 14,000 words per day of twenty-four hours. Finally, on the 10th October, 1856, Professor Morse was enabled to write as follows to the United States: "The doubts are resolved, the difficulties are overcome, success is within our reach, and the great feat of the century must shortly be accomplished." Before October 10th, 1857, it is confidently anticipated that these hopes may be fully realised (great applause).

These festivities having come to an end, the *Agamemnon* set out for Sheerness (after stopping *en route* at Gravesend for the majority of the staff) to adjust compasses. The *Observer* in a report stated :—

When leaving her moorings, opposite Glass & Elliot's Works, the scene was one of considerable interest. Many thousands of persons thronged the river side as far as Greenwich Hospital. In the immediate neighbourhood of the factory a salute was fired as the proud vessel moved away, and a deafening cheer was raised by the assembled crowds. The crew of H.M.S. *Agamemnon* manned the gunwales, and returned the cheer with lusty lungs, while from the stern gallery ladies waved their handkerchiefs, and *savants* forgot for awhile the mysteries of electricity and submarine cable work, as they returned the hearty cheers which reached them from the shore.

The *Agamemnon* was taken in tow by three steam-tugs, one on each side, and a third in front. The tall masts of the giant ship were watched with anxious eagerness till they were lost in the far distance, and her huge hull disappeared amid the numerous bends and windings of the river."

The two ships met at Queenstown, Cork, on July 30th.

Charles Bright at once ran a piece of cable between the ships, which were moored about three-quarters of a mile apart, so as to enable the entire length of 2,500 miles to be tested and worked through.[1] The experiments were continued, by Mr. Whitehouse, for two days, the whole cable proving to be perfect and working through satisfactorily.

---

[1] Owing to the cable having been made in halves at two different factories many miles apart, it will be observed that no proper trials had been possible before.

THE ATLANTIC CABLE

What by that time had become to be known amongst sailors as the "Wire Squadron," sailed. from this *rendezvous* for Valentia Bay on Monday, August 3rd.

After its full strength had been collected at Queenstown, the fleet was composed as follows :—

The U.S. screw-steamer *Niagara* to lay the half of the cable from Valentia Bay, Ireland.

The U.S. paddle-steamer *Susquehanna* (Captain Sands) to attend upon, as consort to, the *Niagara*.

H.M. screw-steamer *Agamemnon*, to lay the half of the cable on the American side.

H.M. paddle-steamer *Leopard* (Captain Wainwright), to attend upon the *Agamemnon*.

H.M. screw-steamer *Cyclops* (Captain Dayman), to go ahead of the steamers and keep the course.

H.M. tender *Advice*, and the steam-tug *Willing Mind*, to assist in landing the cable at Valentia.

Then in Trinity Bay, Newfoundland, the U.S. screw-steamer *Arctic* and the paddle-steamer *Victoria* (chartered by the "Newfoundland" Telegraph Company) were to await the arrival of the fleet, and assist in landing the cable.

Advantage was taken of the passage from Cork to experiment with the paying out machinery, which was found to be perfectly satisfactory.

SECTION IV

The " Wire Squadron" at Valentia

On arrival at Valentia Harbour, on August 4th, the ships were most hospitably welcomed by the Knight of Kerry, Sir Peter Fitzgerald (lord of the soil for many miles round), and his amiable and intellectual family, who had from the commencement taken a keen interest in the project.

Although abutting on the Atlantic, and exposed to the first brunt of all its westerly cyclones—of which we know the severity in England, even after they have broken their force on the coast of the Emerald Isle—the knight's garden on Valentia Island carries one's fancy away to the sunny south of France. Every plant and shrub flourishes, and even the thick hedges are composed of fuchsia plants, some five feet high and several feet thick, covered with flowers from top to bottom.

His Excellency the Earl of Carlisle, Lord Lieutenant of Ireland, with his suite and many friends to the cable, had journeyed from Dublin Castle by special train, and the little corner of Ireland was *en fête* in this "the next parish to America."

At a *déjeuner* banquet, given by the Knight of Kerry, the Lord Lieutenant—who in his earlier days, as Lord Morpeth, had made his mark as a prominent member of the then Government—replying to his host, employed these cogent words of caution as well as of encouragement :—

I believe, as your worthy chairman has already hinted, that I am probably the first Lieutenant of Ireland who ever appeared upon this lovely strand. At all events, no previous Lord Lieutenant can have come amongst you on an occasion like the present.

Amidst all the pride and the stirring hopes which cluster around the work of this week, we ought still to remember that we must speak with the modesty of those who begin, and not of those who close an experiment ; and it behoves us to remember that the pathway to great achievements has frequently to be hewn out amidst risks and difficulties, and that preliminary failure is even the law and condition of ultimate success. Therefore, whatever disappointments may possibly be in store, I must yet insinuate to you that, in a cause like this, it would be criminal to feel discouragement.

In the very design and endeavour to establish the Atlantic telegraph there is almost enough of glory. It is true that if it be only an attempt, there would not be quite enough of profit. I hope that will come too. But there is enough of public spirit, of love for science, for our country, for the human race, almost to suffice of themselves.

However, upon this rocky coast of Ireland, at all events to-day, we will presume upon success. We are about, either at this sundown or at to-morrow's dawn, to establish a new material link between the Old World and the New. Moral links there have been—links of race, links of commerce, links of friendship, links of literature, links of glory—but this, our new link, instead of superseding and supplanting the old ones, is to give a life and an intensity which they never had before.

Highly as I value the reputations of those who have conceived, and those who have contributed to carry out this bright design—and I wish that so many of them had not been unavoidably prevented from being amongst us at this moment—highly as I estimate their reputation, yet do I not compliment them with the idea that they are to efface or dim the glory of that Columbus, who, when the large vessels in the harbour of Cork yesterday weighed their anchors, did so in his own more humble craft on that very day three hundred and sixty-five years ago—it would have been called in Hebrew writ a year of years—and set sail upon his glorious enterprise of discovery. They, I say, will not dim or efface his glory, but they are now giving the last finish and consummation to his work.

Hitherto the inhabitants of the two worlds have associated, perhaps in the chilling atmosphere of distance, with each other—a sort of bowing distance ; but now we can be hand to hand, grasp to grasp, pulse to pulse. The link, which is now to connect us, like the insect in the immortal couplet of our poet :—

" While exquisitely fine,
    Feels at each thread, and lives along the line."

And we may feel, gentlemen of Ireland, of England, and of America, that we may take our stand here upon the extreme rocky edge of our beloved Ireland. We may, as it were, leave behind us in our rear the wars, the strifes, and the bloodshed of the elder Europe, and of the elder Asia. Weak as our agency may be, imperfect as our powers may be, inadequate in strict diplomatic form as our credentials may be ; yet, in the face of the unparalleled circumstances, and as a homage due to science, let us pledge ourselves to eternal peace between the Old World and the New.

During that afternoon the *Agamemnon* and *Niagara* with their consorts hove in sight.

The following morning Charles Bright and his assistants were occupied in completing the arrangements for landing the massive shore end, which, weighing some ten tons to the mile—as already

LANDING THE CABLE AT BALLYCARBERRY, VALENTIA

described—was calculated to withstand damage from any anchorage in the bay. The landing-place which had been finally selected was a little cove known as Ballycarberry, about three miles from Caherciveen, in Valentia Harbour. The two small assistant steamers—*Willing Mind*, a tug with a zeal worthy of her name, and *Advice*, ready not merely with advice but most lusty help—with several other launches and boats, were employed on the operation which commenced at about two o'clock on the afternoon of August 5th, and was thus described in one of the many newspaper reports :—

"Valentia Bay was studded with innumerable small craft decked with the gayest bunting. Small boats flitted hither and thither, their occupants cheering enthusiastically as the work successfully progressed. The cable boats were managed by the sailors of the *Niagara* and the *Susquehanna*. It was a well-designed compliment, and indicative of the future fraternisation of the nations, that the shore rope was arranged to be presented at this side of the Atlantic to the representative of the Queen, by the officers and men of the United States Navy, and that at the other side the British officers and sailors should make a similar presentation to the President of the great republic."

"From the mainland the operations were watched with intense interest. For several hours the Lord Lieutenant stood on the beach, surrounded by his staff and the directors of the railway and telegraph companies, waiting the arrival of the cable. When at length the American sailors jumped through the surge with the hawser to which it was attached, his Excellency was among the first to lay hold of it and pull it lustily to the shore. Indeed, every one present seemed desirous of having a hand in the great work. Never before, perhaps, were there so many willing assistants at the long pull, the strong pull, and the pull all together."

"At half-past seven o'clock the cable was hauled on shore at Ballycarberry Strand, and formal presentation was made of it to the Lord Lieutenant by Commander Pennock, of the *Niagara*, his Excellency expressing a hope that the work so well begun would be carried to a satisfactory completion." [1]

After the Rev. Mr. Day, the vicar of the parish, had offered a prayer for the success of the undertaking, the Lord Lieutenant closed the proceedings with the following inspiriting remarks :—

My American, English and Irish friends, I feel at such a moment as this, that no language of mine can be becoming except that of prayer and praise.

However, it is allowable to any human lips, though they have not specially qualified for the office, to raise the ascription of "Glory to God in the highest ; on earth peace, good-will to men." That I believe is the spirit in which this great work has been undertaken, and it is this reflection that encourages me to feel hope in its final success.

I believe that the great work now so happily begun will accomplish many great and noble purposes for trade, for national policy, and for the Empire.

But there is only one view in which I will present it to those whom I have the pleasure to address. You are

---

[1] *The Liverpool Daily Post*, August 7th, 1857.

aware, you must know—some of you from your own experience—that many of your dear friends and near relatives have left their native land to receive hospitable shelter in America. Well then, I do not expect that all of you can understand the wondrous mechanism by which this great undertaking is to be carried on. But this, I think, you will all of you understand. If you wished to communicate some piece of intelligence straightway to your relatives across the wide world of waters—if you wished to tell those whom you knew it would interest in their heart of hearts of a birth, or a marriage, or, alas! a death amongst you—the little cord which we have now hauled up to the shore will impart those tidings quick as the flash of lightning!

Let us indeed hope—let us pray—that the hopes of those who have set on foot this great design, may be rewarded by its entire success. And let us hope further, that this Atlantic Cable will, in all future time, serve as an emblem of that strong cord of love, which I trust will always unite the British Islands to the great continent of America. And you will join me in my fervent wish that the Giver of all good, who has enabled some of his servants to discern so much of the working of the mighty laws by which He fills the Universe—will further so bless this wonderful work, as to make it evermore to serve the high purpose of the good of man, and tend to His great glory.

And now, my friends, as there can be no project or undertaking which ought not to receive the approbation and applause of the people, will you join with me in giving three hearty cheers for it? (loud cheering).

Three cheers are not enough for me, they are what we give on common occasions, and as it is for the success of the Atlantic Telegraph Cable, I must have at least one dozen cheers! (loud and prolonged cheering).

THE EARL OF CARLISLE (LORD LIEUTENANT OF IRELAND) MAKING A SPEECH AT THE START OF THE EXPEDITION

THE ATLANTIC CABLE

The work connected with the landing of the shore end was not actually completed till sunset; so, as it was too late to proceed on their journey, the ships remained at anchor in the bay till daybreak. That night there was a grand ball at the little village of Knightstown, and the day dawn caught the merry-makers still engaged in their festivities. A bonfire of peat, piled up as high as a good-sized two-story house, sent its ruddy and cheerful light far out into the darkness, brightening up the black crevices in the frowning rocks, and throwing a glow on the faces of the light-hearted peasantry that gathered around in a huge circle.

SECTION V

Laying the First Ocean Cable

Charles Bright, with his chief assistants, Messrs. Canning, Woodhouse and Clifford, as well as Mr. Kell (whose duty it was to look after the cable stores and men), had taken up their quarters on board the *Niagara*, besides Bright's brother-in-law, Mr. Robert John Taylor, who accompanied the expedition as a visitor and looker-on.[1] Here were also Mr. Field and Professor Morse.[2]

[1] Thus, in after years when asked whether he had ever crossed the Atlantic Mr. Taylor was wont to puzzle his enquirers by the response "half-way!"

[2] The latter, who besides being electrician to the

Mr. Webb was quartered on the *Agamemnon*, together with Professor Thomson and Mr. H. A. Moriarty[1] as Navigating Master—specially detailed by the Admiralty, on account of his skill, for this work.

Mr. C. V. de Sauty, a gentleman of considerable practical experience, was placed in charge of the electrical arrangements on board the *Niagara*, and subservient to the wishes of Mr. Whitehouse from shore. Mr. J. C. Laws was also there. Other assistant electricians connected with the enterprise, on board either of the ships (or else ashore), to work for the Company under the instructions of Mr. Whitehouse or Prof. Thomson, were Mr. E. G. Bartholomew, on the *Agamemnon* (Mr. Thomson's right-hand man); Mr. F. Lambert, who afterwards became a prominent electrician and member of Messrs. Bright & Clark's staff; Mr. H. A. C. Saunders; Mr. Benjamin Smith, now an authority on all electrical matters connected with cables; Mr. Richard Collett, and Mr. Charles Gerhardi. Mr. Whitehouse was not able to go out on the expedition for reasons of health. A gentleman closely

New York and Newfoundland Telegraph Company, held also an honorary watching brief on behalf of the United States Government, had unfortunately to retire to his berth as soon as the elements began to assert themselves, and remained there more or less continuously throughout the expedition.

[1] Afterwards Staff-Commander Moriarty, C.B.

associated with the latter in most of his electrica-
researches was Mr. Samuel Phillips, sen. Mr. Saunders has since taken a prominent part in the extension of submarine telegraphs as chief electrician to
the "Eastern" and "Eastern Extension" Companies.

MR. MORIARTY, R.N. (NOW CAPTAIN H. A. MORIARTY, C.B.)

Mr. Gerhardi has for many years represented the
" Direct Spanish " Company as its manager, whilst
Mr. Collett is now the worthy secretary of the
" Brazilian Submarine " Company.

In addition, the expedition was accompanied by
Mr. Nicholas Woods, representing *The Times* news-

paper. Mr. Woods wrote a number of articles[1] which were much appreciated by readers of that journal. These articles gave detailed accounts of the expedition as it progressed, somewhat in the form of a narrative. His brother, Mr. J. E. Tennison Woods, was on board the *Agamemnon* in a like capacity for the *Daily News*. Similarly, Mr. John Mullaly[2] was aboard U.S S. *Niagara*, reporting on behalf of the *New York Herald*. Thus it will be seen that the Press was well represented.

The ships got under way at an early hour on the morning following the landing of the shore end. Paying out was commenced from the forepart of the *Niagara*, and as the distance from that to the stern was considerable, a number of men were stationed at intervals, like sentinels, to see that every foot of the line reached its destination in safety. The machinery did not seem at first to take kindly to its work, giving vent to many ominous groans. After five miles had been disgorged in safety, the bulky line caught in some of the apparatus and parted. The good ship at once put back, and the cable was under-run by the *Willing Mind* (with boats) the whole distance from the shore—a tedious and hard task as may be imagined. At

---

[1] These have been largely drawn upon in our pages.

[2] This gentleman afterwards published a book on the subject, treated from a strictly American standpoint.

length the end was lifted out of the water and spliced to the gigantic coil on board; and as it dropped safely to the bottom of the sea, the mighty ship steamed ahead once more.

At first she moved very slowly, not more than two miles an hour, to avoid the danger of another accident, but the feeling that they were at last away was in itself a relief. The ships were all in sight, and so near that they could hear each other's bells. The *Niagara*, as if knowing she was bound for the land out of whose forests she came, bowed her head proudly to the waves, her prow directed towards her native shores.

"Slowly passed the hours of that day," in Mr. Henry Field's words;[1] "but all went well, and the ships were moving out into the broad Atlantic. At length the sun went down in the west, and stars came out on the face of the deep. But no man slept. A thousand eyes were watching a great experiment, including those who had a personal interest in the issue.

"All through that night, and through the anxious days and nights that followed, there was a feeling in the heart of every soul on board, as if some dear friend were at the turning point of death, and they were watching beside him. There was a strange unnatural silence in the ship. Men paced the deck with soft and muffled

[1] Brother of Mr. Cyrus Field, who later on wrote an animating description of the enterprise.

tread, speaking only in whispers, as if a loud or heavy footfall might snap the vital cord. So much had they grown to feel for the enterprise, that the cable seemed to them like a human creature, on whose fate they themselves hung, as if it were to decide their own destiny.

"There are some who will never forget that first night at sea. Perhaps the reaction from the excitement on shore made the impression the deeper. There are moments in life when everything comes back to us. What memories cropped up in those long night hours! How many on board that ship, as they stood on the deck and watched that mysterious cord disappearing in the darkness, thought of homes beyond the sea, of absent ones, of the distant and of the dead.

"But no musings turn them from the work in hand. There are vigilant eyes on deck—Mr. Bright, the engineer-in-chief, is there; also, in turn, Mr. Woodhouse and Mr. Canning, his chief assistants. . . . The paying-out machinery does its work, and though it makes a constant rumble in the ship, that dull heavy sound is music in their ears, as it tells them that all is well. If one should drop to sleep, and wake up at night, he has only to hear the sound of 'the old coffee-mill' and his fears are relieved, and he goes to sleep again."

Saturday was a day of beautiful weather. The ships were getting further away from land, and began to steam ahead at the rate of four and five knots. The cable was paid out at a speed a little faster than the ship, to allow for inequalities of surface on the bottom of the sea. While it was thus going overboard, communication was kept up constantly with the land.

To quote Mr. Henry Field again :—

" Every moment the current was passing between ship and shore. The communication was as perfect as between Liverpool and London, or Boston and New York. Not only did the electricians telegraph back to Valentia the progress they were making, but the officers on board sent messages to their friends in America to go out by the steamers from Liverpool. The heavens seemed to smile on them that day The coils came up from below the deck without a kink, and unwinding themselves easily, passed over the stern into the sea.

" All Sunday the same favouring fortune continued ; and when the officers who could be spared from the deck met in the cabin, and Captain Hudson read the service, it was with subdued voices and grateful hearts that they responded to the prayers to ' Him who spreadeth out the heavens and ruleth the raging of the sea.'

" On Monday they were over two hundred miles at sea. They had got far beyond the shallow waters off the coast. They had passed over the submarine mountain that figures on the

charts of Dayman and Berryman, and where Mr.
Bright's log gives a descent from five hundred
and fifty to seventeen hundred and fifty fathoms
within eight miles. Then they came to the deeper
waters of the Atlantic where the cable sank to
the awful depth of two thousand fathoms! Still
the iron cord buried itself in the waves, and every
instant the flash of light in the darkened telegraph
room told of the passage of the electric current.

"Everything went well till 3.45 p.m. on the
fourth day out, the 11th August, when the cable
snapped after 380 miles had been laid, owing to
mismanagement on the part of the mechanic at
the brakes."

Thus, the familiar thin line which had been
streaming out from the *Niagara* for six days was
no longer to be seen by the accompanying vessels.

One who was present wrote :—

"The unbidden tear started to many a manly eye.
The interest taken in the enterprise by officers and
men alike exceeded anything ever seen, and there
is no wonder that there should have been so much
emotion on the occasion of the accident."

The following report from Charles Bright gives
the details of the expedition up to the time of
this regrettable occurrence :—[1]

[1] Report to the Directors of the Atlantic Telegraph
Company, August, 1857 :—
"After leaving Valentia on the evening of the 7th

inst. the paying out of the cable from the *Niagara* progressed most satisfactorily until immediately before the mishap.

"At the junction between the shore and the smaller cable, about eight miles from the starting-point, it was necessary to stop to renew the splice. This was successfully effected, and the end of the heavier cable lowered by a hawser until it reached the bottom, two buoys being attached at a short distance apart to mark the place of union.

"By noon of the eighth we had paid out forty miles of cable, including the heavy shore end. Our exact position at the time was in lat. 50° 59′ 36″·N., long. 11° 19′ 15″ W., and the depth of the water according to the soundings taken by the *Cyclops* — whose course we nearly followed—ninety fathoms. Up to 4 p.m. on that day the egress of the cable had been regulated by the power necessary to keep the machinery in motion at a slightly higher rate than that of the ship ; but as the water deepened, it was necessary to place some further restraint upon the cable by applying pressure to the friction drums in connection with the paying-out sheaves. By midnight, eighty-five miles had been safely laid, the depth of the water being then a little more than 200 fathoms.

"At eight o'clock in the morning of the 9th, we had exhausted the deck coil in the after part of the ship, having paid out 120 miles. The change to the coil between decks forward was safely made. By noon we had laid 136 miles of cable, the *Niagara* having reached lat. 52° 11′ 40″ N., long. 13° 0′ 20″ W., and the depth of the water having increased to 410 fathoms. In the evening the speed of the vessel was raised to five knots. I had previously kept down the rate at from three to four knots for the small cable, and two for the heavy end next the shore, wishing to get the men and machinery well at work prior to attaining the speed which I had intended making. By midnight, 189 miles of cable had been laid.

"At four o'clock in the morning of the 10th the depth began to increase rapidly, from 550 to 1,750 fathoms in a distance of eight miles. Up to this time a strain of 7 cwt. sufficed to keep the rate of the cable near enough to that of the ship; but as the water deepened the proportionate speed of the cable advanced, and it was necessary to augment the pressure by degrees until at a depth of 1,700 fathoms, the indicator showed a strain of 15 cwt. while the cable and the ship were running $5\frac{1}{2}$ and 5 knots respectively.

"At noon on the 10th we had paid out 255 miles of cable—the vessel having made 214 miles from the shore —being then in lat. 52° 27′ 50″ N., long. 16° 15′ W. At this time we experienced an increasing swell, followed later in the day by a strong breeze.

"From this period, having reached 2,000 fathoms of water, it was necessary to increase the strain by a ton, by which the rate of the cable was maintained in due proportion to that of the ship. At six o'clock in the evening some difficulty arose through the cable getting out of the sheaves of the paying-out machine, owing to the pitch and tar hardening in the grove and a splice of large dimensions passing over them. This was rectified by fixing additional guards, and softening the tar with oil. It was necessary to bring up the ship, holding the cable by stoppers, until it was again properly disposed around the pullies. Some importance is due to this event, as showing that it is possible to 'lay to' in deep water without continuing to pay out the cable, a point upon which doubts have frequently been expressed.

THE ATLANTIC CABLE

"Shortly after this the speed of the cable gained considerably on that of the ship, and up to nine o'clock, while the rate of the latter was about three knots, by the log, the cable was running out from five and a half to five and three quarter knots.

"The strain was then raised to 25 cwt., but the wind and the sea increasing, and a current at the same time carrying the cable at an angle from the direct line of the ship's course, it was found insufficient to check the cable, which was at midnight making two and a half knots above the speed of the ship, and sometimes imperilling the safe uncoiling in the hold.

"The retarding force was therefore increased at two o'clock to an amount equivalent to 30 cwt., and then again, in consequence of the speed continuing to be more than it would be prudent to permit, to 35 cwt. By this the rate of the cable was brought to a little short of five knots, at which it continued steadily until 3.45, when it parted, the length paid out at the time being 380 miles.

"I had, up to this, attended personally to the regulation of the brakes; but finding that all was going on well, and it being necessary that I should be temporarily away from the machine—to ascertain the rate of the ship, to see how the cable was coming out of the hold, and also to visit the electricians' room—the machine was for the moment left in charge of a mechanic who had been engaged from the first in its construction and fitting, and was acquainted with its operation.

"In proceeding towards the fore part of the ship I heard the machine stop; I immediately called out to relieve the brakes; but when I reached the spot, the cable was broken. On examining the machine, which was otherwise in perfect order, I found that the brakes had *not* been released ; and to this, or to the hand wheel of the brake being turned the wrong way, may be attributed the stoppage, and consequent fracture, of the cable.

" When the rate of the wheels grew slower, as the ship dropped her stern in the swell, the brake should have been eased. This had been done regularly whenever an unusually sudden descent of the ship temporarily withdrew the pressure from the cable in the sea. But owing to our entering the deep water the previous morning and having all hands ready for any emergency that might occur there, the chief part of my staff had been compelled to give in at night through sheer exhaustion ; and hence, being short-handed, I was obliged for the time to leave the machine without, as it proves, sufficient intelligence to control it.

" I perceive that on the next occasion it will be needful, from the wearing and anxious nature of the work, to have three separate relays of staff, and to employ, for attention to the brakes, a higher degree of mechanical skill.

" The origin of the accident was, no doubt, the amount of retarding strain put upon the cable ; but had the machine been properly manipulated at the time, it could not possibly have taken place.

" For three days, in shallow and deep water, as well as in rapid transition from one to the other, nothing could be more perfect than the working of the cable machinery. It had been made extra heavy with a view to recovery work. However, it performed its duty so smoothly and efficiently in the smaller depths—where the weight of the cable had less ability to overcome its friction and resistance—that it can scarcely be said to be too heavy for paying out in deep water, where it was necessary, from the increased weight of cable, to restrain its rapid motion, by applying to it a considerable degree of additional friction.

Its action was most complete, and all parts worked well together.

"I see how the gear can be improved, by a modification in the form of sheave, by an addition to the arrangement for adjusting the brakes, and some other alterations; but with proper management, without any change whatever, I am confident that the whole length of cable might have been safely laid by it. And it must be remembered—as a test of the work which it has done—that unfortunate as this termination to the expedition is, the longest length of cable ever laid has been paid out by it, and that in the deepest water yet passed over.

"After the accident had occurred, soundings were taken by Lieutenant Dayman from the *Cyclops*, and the depth found to be 2,000 fathoms.

"It will be remembered that some importance was attached to the cable on board the *Niagara* and *Agamemnon* being manufactured in opposite lays. I thought this a favourable opportunity to show that practically the difference was not of consequence in effecting the junction in mid-ocean. We therefore made a splice between the two vessels. This was then lowered in a heavy sea, after which several miles were paid out without difficulty.

"I requested the commanders of the several vessels to proceed to Plymouth, as the docks there afford better facilities than any other port for landing the cable, should it be necessary to do so.

"The whole of the cable remaining on board has been carefully tested and inspected, and found to be in as perfect condition as when it left the works at Greenwich and Birkenhead, respectively.

"One important point presses for your consideration at an early period. A large portion of cable, already laid may be recovered at a comparatively small expense. I append an estimate of the cost, and shall be glad to receive your authority to proceed with this work.

"I do not perceive in our present position any reason for discouragement ; but I have, on the contrary, a greater confidence than ever in the undertaking.

"It has been proved beyond a doubt that no obstacle exists to prevent our ultimate success, and I see clearly how every difficulty which has presented itself in this voyage can be effectually dealt with in the next.

"The cable has been laid at the expected rate in the great depths ; its electric working through the entire length has been satisfactorily accomplished ; while the portion laid, actually improved in efficiency by being submerged—from the low temperature of the water and the increased close texture of gutta percha thereby effected.

"Mechanically speaking, the structure of the cable has answered every expectation that I had formed of it Its weight in water is so adjusted to the depth, that strain is within a manageable scope : while the effects of the undercurrents upon its surface, prove how dangerous it would be to lay a much lighter rope, which would, by the greater time occupied in sinking, expose an increased surface to their power, besides its descent being at an angle such as would not provide for good laying at the bottom.

"I have the honour to remain, Gentlemen,
"Yours very faithfully,
"CHARLES T. BRIGHT,
"*Engineer-in-Chief.*
"*To the Directors of the Atlantic Telegraph Company.*"

This Report [1] was afterwards sent by the Secretary

[1] Followed some weeks later by a more complete report, including an account of subsequent operations, besides the engineer's recommendations for the future. This further report of Bright's is given in the Appendices.

of the Company to *The Times* newspaper, together with the following letter :—

THE ATLANTIC TELEGRAPH.

SIR,—

I beg to annex an official copy of the Chief Engineer's Report of the operations on board of the United States frigate *Niagara* up to the date of the accident to the Atlantic Cable.

Yours, etc.,

GEORGE SAWARD, *Secretary.*

22, Old Broad Street, London, August 19th, 1857.

SECTION VI

Preparations for another Attempt

This untoward interruption to the expedition was naturally a cause of great disappointment to all connected with the undertaking, as there was not enough cable left to complete the work, nor was there time to get more made and stowed on board to renew the attempt before the season would be too far advanced.

Still, much experience had been gained, and there were many points of encouragement in Charles Bright's Report. Those immediately concerned in the great enterprise were, despite their heavy disappointment, in the end undaunted.

The squadron proceeded to Plymouth to unload

the cable into tanks at Keyham Dockyard,[1] chiefly because some of the ships could not be spared by their respective Governments till the following year.

The insulation was carefully tested by Professor Thomson and Mr. Whitehouse, who found that the copper wire had forced its way through the gutta-percha at several points—probably owing to the repeated coiling and uncoiling—the manufacture of gutta-percha at the proper temperature not being then understood as it is now, and still less that of gutta-percha core. These defects were carefully repaired.

On arrival at Plymouth, Charles Bright journeyed to the head-quarters of the Atlantic Telegraph Company in London, and soon after, arrangements having been made for the future, he penned these letters :—

Sept. 15, 1857.

MY DEAR SIR,—

I have seen your letter to Mr. Saward. I shall run down to Plymouth to-morrow night.

The naval authorities on board, and in the Dockyard at Plymouth, must not expect us to begin uncoiling at once. It will take some time, when we get leave to unload, to prepare a suitable tank to receive the cable so that it may be properly protected during the winter.

[1] Forming a part of the present Devonport Dockyard.

THE ATLANTIC CABLE

Will you let Captn. Hudson know this so that he may not be disappointed?

I am, yours very truly,

CHARLES T. BRIGHT.

WILDMAN WHITEHOUSE, Esq.,
Royal Hotel, Plymouth.

CAPT. KELL,   *Sept. 15, 1857.*
    Plymouth.

DEAR SIR,—

I shall run down to Plymouth to-morrow night to arrange about the uncoiling.

Do not let it be thought on board that we can begin at once, for even when we have leave to begin, the platform, etc., will take some time making. It ought to be caulked round the sides as well, so that we can let water in where the cable lies.

Yours very truly,

CHARLES T. BRIGHT.

During Charles Bright's absence in London, certain matters were also attended to for him at Devonport by Mr. Temple, whilst the entire control of the ultimate unshipment of the cable rested with Mr. Canning and Mr. Woodhouse. On being discharged from the ships, the cable was passed through a composition of tar, pitch, linseed-oil, and bees-wax (as a precaution against oxidation); and was coiled in compact circles in four large roofed tanks specially constructed for the purpose, with a view to storing the cable ashore until the following summer, when the undertaking was to be resumed —at least so many hoped.

In the middle of October Bright proceeded to Valentia, accompanied by Mr. Clifford, in a small paddle-steamer, with the object of picking up some of the cable near here. After experiencing a series of gales, over fifty miles of the main cable were recovered and the shore end buoyed ready for splicing on to in the coming year. Whilst engaged

PICKING UP THE CABLE

in the above work the subject of our biography penned the following to his wife which serves to describes the operation and the apparatus employed :—

VALENTIA, *October* 24, 1857.

I send you a gift from H. C——, a view from our window at the inn here. The steamer to the right is the *Leipsig*. The pier is the breakwater of Valentia

Harbour. The queer looking thing to the left is an apparatus I have fitted up for under-running the cable. It is composed of two very large long iron buoys fixed together like a twin ship with a platform of timber over it.[1] On this, at each end, is a saddle with a deep groove for the cable to run in. The cable being on the near shore, it is towed along. When near the end of the heavy cable I shall take it off, cut the cable, buoy the heavy end, and begin winding up the small one as we go on.

This first expedition had opened the eyes of the investing public to the vastness of the undertaking, and led many to doubt who did not doubt before. Some even began to look upon it as a romantic adventure of the sea, rather than as a serious commercial undertaking.

As Henry Field reminds us :—" This decline of popular faith was felt as soon as there was a call for more money."

The loss of 335 miles of cable with the postponement of the expedition to another year, was equivalent to a loss of £100,000. To make this good, the capital of the company had to be increased, and this new capital was not so readily obtainable. The projectors found that it was easy to go with the current of popular enthusiasm, but very hard to stem a growing tide of popular distrust. There were many who remembered Lord Carlisle's words of warning, previous to the start from Valentia.

[1] See accompanying illustration.

And it must also be remembered that, from the very first, that section of the public which looked with distrust upon the idea of an Atlantic Telegraph, was far in excess of that which did not; indeed, the opposition encountered was much on a par with the great popular prejudice which George Stephenson had to overcome when projecting his great Railway schemes.

RE-SHIPMENT OF THE CABLE ABOARD H.M.S. *AGAMEMNON* AND U.S.N.S. *NIAGARA* IN KEYHAM BASIN

But whatever the depression at the untimely termination of the first expedition, it did not interfere with renewed and vigorous efforts to prepare for a second. In the end the appeal to the shareholders for more money was responded to, and the directors were enabled to give orders for the manufacture of 700 miles of new cable of the same description, to make up for what had been lost, and to provide a surplus against all contingencies. Thus, 3,000 nautical miles in all were eventually shipped this time, instead of 2,500 miles.

Whilst the cable was being coiled on the *Agamemnon* and *Niagara*, an opportunity was accorded Professor D. E. Hughes, F.R.S. (who has since done so much for telegraphy and telephony), to experiment with his beautiful type-printing apparatus, at that time largely employed in the United States, and in Europe. He found that he was able to write through the entire length at the rate of some thirty, or more, letters per minute.[1]

A committee was arranged to confer with Charles Bright as to the machinery, consisting of Mr. Thomas Lloyd, the chief of the Steam Department of Her Majesty's Navy; Mr. John Penn,[2] of Greenwich; and Mr. Joshua Field,

---

[1] Later on, trials were made by Mr. Thomas Edison with his high speed instrument on a subsequent Atlantic cable—that of 1873; but these experiments did not have a successful issue, and Mr. Edison, with a candour seldom indulged in by inventors (except those who, like Mr. Edison, can afford it), expressed himself that "he was not prepared for this darned induction, and he would go back to America to think about it."

[2] Later on Mr. Penn became one of Charles Bright's great local political supporters at the Greenwich Election in 1865.

F.R.S., of Maudslay, Son & Field. Mr. W. E. Everett, U.S.N., was also consulted later. As the chief (ship's) engineer of the *Niagara*, on the late expedition, Mr. Everett had acquired a good deal of information from seeing the working of the apparatus on board. He also joined in approving Charles Bright's suggested alterations. This gentleman had to return to America with his ship; but on again arriving in London, on January 18th of the following year, he had the satisfaction of attesting to the sterling qualities of the machine devised, adopted, and constructed in his absence, as well as in partly superintending the setting up of it aboard the ships.

The above committee reported [1] "we consider the paying out sheaves require no alterations except those suggested by Mr. Bright in a memorandum he was good enough to place in our hands."

Quite independently, however, Charles Bright had decided that the checking gear, or brake, should not be left in the power of any person in charge to jamb the machine. Subsequently a very apportune invention of Mr. J. G. Appold, F.R.S., was suggested. It consisted of a brake so arranged that a lever exercised a uniform holding power in exact proportion to the weights attached to it, and while capable of being *released* by a hand-wheel, it could not be tightened. This clever appliance had been introduced in connection

[1] 17th September, 1857.

with the crank apparatus in gaols, so as to regulate the amount of labour in proportion to the strength of the prisoner.

The above invention was especially adapted to the exigencies of cable work on this occasion, for the first time, by Mr. C. E. Amos, M.Inst.C.E.,[1] and Charles Bright, who also consulted with Mr. Clifford in the matter, as well as with Mr. Canning.

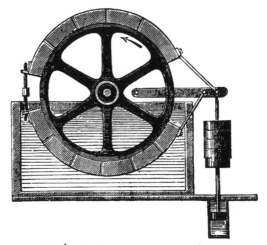

AMOS AND BRIGHT'S MODIFICATION OF APPOLD'S FRICTION BRAKE

The great feature about it was that it provided for automatic release of the brake, upon the strain exceeding that intended. Thus only a maximum agreed strain could be applied, this being regulated from time to time by weights, according to depth of water and consequent weight of cable being paid out.

[1] Of the famous engineering firm of Easton & Amos.

In passing from the hold to the stern of the laying vessel, the cable is taken round a drum.[1] Attached to the axle of the drum[2] is a wheel fitted with an iron friction-strap (to which are fixed blocks of hard wood), capable of exerting a given retarding power, varying with the weights hung on to the lever which tightens the strap. When the friction becomes great, the wheels have an increased tendency to carry the wooden blocks

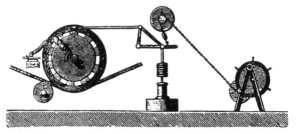

AMOS AND BRIGHT'S FRICTION-BRAKE APPARATUS

round with them : thus the lever bars are deflected from the vertical line and the iron band opened sufficiently to lessen the brake power. Hence this apparatus may be said to be partially self regulating in its action, to the extent of avoiding an excess retarding force.

[1] In the actual apparatus for the laying of the 1858 cable, there were *two* drums, each having two brake wheels attached to their axes, as will be seen in the opposite plate and accompanying description.

[2] This drum is carried round by the weight of the axle as the ship moves onwards.

Charles Bright also devised a dynamometer apparatus — for indicating and controlling of the strain during paying out—which was a vast improvement on that embodied in the previous machines.

The working connections of the friction-brake and hand-wheel, referred to, may be seen here.

A more complete notion of it, however, as well as of the entire paying-out gear (with Bright's dynamometer), as ultimately adopted for the next expedition, is best obtained from the accompanying plate.

The working of the machine was as follows :—[1]

Between the stern of the vessel and the machine the cable was bent somewhat out of the straight

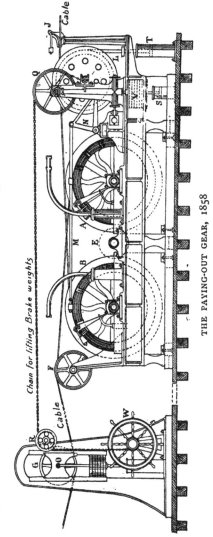

THE PAYING-OUT GEAR, 1858

---

[1] *The Electric Telegraph*, by E. B. Bright, M.Inst.C.E.

line by being led under the grooved wheel of the dynamo-meter. This wheel had a weight attached to it, and could be moved up or down in an iron frame. If the strain upon the cable was small, the wheel would bend the cable downwards, and its index would show a low degree of pressure; but whenever the strain increased, the cable, in straightening itself, would at once lift the dynamometer wheel with the indicator attached to it, which showed the pressure in cwts. and tons. The amount of strain with a given weight upon the wheel was determined by experiments, and a hand wheel in connection with the levers of the paying-out machine was placed immediately opposite the dynamometer: so that, directly the indicator showed strain increasing, the person in charge could at once, by turning the hand-wheel, lift up the weights that tightened the friction straps, and so let the cable run freely through the paying-out machine. Although, therefore, the strain could be *reduced*—or entirely withdrawn—in a moment, it could not be *increased* by the man at the wheel.

The dynamometer principle of Charles Bright here introduced has been universally adopted in the laying of all subsequent submarine cables.

From an examination of the illustration, it will be observed that—

" The four U-sheaves of the former expedition were replaced by two grooved drums [1] (about $5\frac{1}{2}$ feet in dia-

---

[1] Though theoretically perfect, the U-sheaves had been found to introduce practical difficulties by the accumulation of tar and compound when paying out a long continuous length of cable. In the new machine

meter) placed tandem fashion outside their bearings
one in front of the other[1] Then two smooth brake-
wheels, placed side by side, were "keyed" on to each
drum-shaft, and partially surrounded with wooden blocks
held in place by strong brake-straps of sheet-iron, so
arranged with compensating levers that any excess of
strain beyond that which is desired, releases the pressure.[2] The cable, in coming from the tanks, passed
under a lightly-weighted "jockey" pulley. This arrangement, whilst leading the line on to the drums, at the
same time checked it slightly.[3] From here it was

---

each drum had tour grooves in which tne cable rested
without getting too deeply embedded. The grooves were
intended to act mainly as a *guide*, or lead, in fact.

[1] Grooved drums had, in 1857, been applied by Charles
Bright for picking up. In fact, this amounted to the
"hauling back" machine of tne previous expedition being
transformed into a paying-out machine, with the principle
of Appold's brake turned to account for cable-laying.

[2] This entire apparatus — simplified as regards the
brake—has since been universally adopted for submarine
cable-work, with the exception that a single flanked
drum, fitted with a sort of plough, skid, or knife-edge,
to guide or "fleet" the incoming turn of cable correctly
on to the drum is used in place of the grooved sheaves.
The latter were first suggested by a device of Professor
W. J. M. Rankine, and Mr. John Thomson (a relation
of Professor Thomson) who had served Mr. Newall, and
afterwards Mr. Brett, in laying various cables.

[3] The first use here of the above "jockey" system
as well as the first application of the principle of
Appold's brake, was later on alluded to by Mr. Elliot

guided by a grooved pulley, or V-sheave, along the tops of both drums, then three times round them, and hence over another V-sheave, and on to the dynamometer. From this the cable was led over a second pulley, and so into the sea by the stern-sheaves as may be seen on all telegraph ships in the present day." [1]

The construction of this improved apparatus was carried out by Messrs. Easton & Amos at their works in Southwark. The working out of the details was greatly assisted by Mr. W. E. Everett, U.S.N., who, in the spring of 1858, joined the cable-laying staff,[2] from which Mr. Webb had previously retired to enter upon another engagement. Mr. Henry Clifford, who as a mechanical engineer was an expert in machinery, also attended closely to the manufacture of the gear. Messrs. Field, Penn, and Lloyd also advised with Bright and his assistants, from time to time, regarding the apparatus.

Suggestions were also made by Professor Thom-

(afterwards Sir George Elliot, Bart., M.P.), in the course of a paper read at the Institution of Mechanical Engineers.

[1] *Submarine Telegraphs*, by Charles Bright, F.R.S.E.

[2] Mr. Everett had been chief engineer on U.S. frigate *Niagara* during the previous expedition. Besides being a fellow countryman, he was a great *protégé* of Mr. Cyrus Field, and in this way became more closely identified with the future operations as regards the cable.

THE ATLANTIC CABLE

son, who had in the previous October contributed some valuable articles to *The Engineer*—at that time the only technical journal of its kind—concerning the mathematical theory of the form assumed by a cable while sinking, and the strains to which it is subjected under various conditions.

Again, just as in 1857, Mr. F. R. Window had read a paper before the Institution of Civil Engineers—the first paper on the subject of Submarine Telegraphy, but of a purely record character.[1] So, in 1858, two able calculating engineers, Mr. T. A. Longridge, M.Inst.C.E., and Mr. C. H. Brooks read another.[2] This was of a mathematical and almost entirely theoretical nature—regarding, the resistance to a rope introduced by skin friction in passing through water.[3] These

[1] See *Inst. C. E., Mins. Proc.*, vol. xvi.

[2] *Proc. Inst. C. E.*, vol. xvii.

Young Bright led off the discussion on the above paper—being naturally looked to for authentic information regarding the Atlantic cable—and his remarks thereon are reproduced in full in the Appendix VIII. to this volume.

[3] They considered the case similar to that of air, or water, flowing through pipes. The views here enunciated by Messrs. Brooks and Longridge were somewhat warmly called into question about twenty years afterwards by the late Dr. Werner von Siemens, in the course of a contribution to the Society of Telegraph Engineers. Dr. Siemens averred that the longitudinal friction so set up varied directly as the velocity rather than as the square of the

gentlemen also asserted, from their mathematica deductions, that:[1]—The result of a stoppage of the paying-out apparatus, in a depth of 2,000 fathoms, whilst the vessel was proceeding at the rate of six feet per second, would be to bring a strain on the cable amounting to over seven tons, while its strength was only about half that. It was, however, actually in evidence that Charles Bright had stopped paying out, during the last expedition, when in very nearly that depth for some length of time while clearing tar off the brake machine.

The following week the "Civils" engaged once more in submarine cable talk. This time an eminently practical discourse was furnished by Mr. F. C. Webb;[2] and here was the occasion on which Professor Airy,[3] the then Astronomer Royal, expressed himself very decidedly that (1st) "it was a mathematical impossibility to submerge the cable successfully at so great a depth in safety," and (2nd) that, "if it were possible, no signals could be transmitted through so great a length."

Professor Airy stated in addition that, "When

velocity. Later again, Mr. Charles Hockin, M.A. (a Cambridge Senior Wrangler), arrived at a computation about mid-way between these. In practice, however, it is neither easy to ascertain the truth, nor is it a matter of great moment.

[1] *Proceedings Inst. C. E.*

[2] *Inst. C. E., Proc.*, vol. xvii.

[3] Afterwards Sir George Biddell Airy, K.C.B., F.R.S.

a cable was paid out at the angle of 10° (with the horizon), there was a strain upon it of nearly sixty-six times the minimum tension, or sixty-six times the depth of the sea at that place. When this was considered, it seemed to him, that in all the annals of engineering there was not another instance in which danger was incurred so needlessly. . . . The angle at which it should be paid out should never be less than 45° with the horizon." He also supported his theory by a series of computations, but altogether omitted to take into account the fact that a cable ship merely moves quickly, as it were, from under the cable it carries, and that the faster the paying out can be carried on, the closer is the angle of the cable to the horizon—10° to 15° being very customary.

Charles Bright's speech at this meeting is given in full in the Appendices. He took the opportunity here of correcting some of the wildly erroneous notions that had obtained currency and given rise to false conclusions.

From the very outset of the project, and as soon as he was appointed engineer, young Bright had had to deal with unfledged amateurs in the art! As the "Jack-in-office" all were ready to pounce upon him : thus he was subjected to all manner of wild suggestions regarding the laying of cables. transmitted to him officially through the secretary of the Company, Mr. Saward. These—which, with our present lights do indeed seem ludicrous—had to be politely replied to notwithstanding extreme pressure of work. Many, be it added, emanated from men of highly scientific attainments.

From the start, Bright was also continually approached by a number of inventors with various crude inventions. Each considered that his particular plan should be adopted, and—what was to him more to the point—paid for by the Atlantic Company, which seemed to be regarded as a kind of milch cow. The projected cable, and the formation of such a company, was enough to stimulate and excite the brains of many a sanguine inventor. Unhappily our engineer had to be the recipient of their lucubrations, and, in some cases, plenty of abuse into the bargain, if he ventured to argue against their suggestions.

In the course of an interesting article in the *Edinburgh Review*, for December, 1866, Prof. A. Fleeming Jenkin, F.R.S., gave a full account of all these unpractical notions. It is both amusing and sad to read of them. A few of the more prominent ideas may be suitably adverted to here.

Perhaps the most frequent was in connection with the fallacy, that as the water increased in depth, therefore at a given point in deep water, a long way off the bottom, the cable would be held in suspense. This was a very common delusion at the time. Of course the pressure increases with the depth, on all sides of a cable (or anything else), in its descent through the sea; but as, practically, everything on earth is more compressible than water, it is obvious that the iron wire, yarn, gutta percha, and copper conductor, forming the cable, must be more and more compressed as they descend. Thus the cable constantly increases its density, or specific gravity, in going down; while the equal bulk of the water surrounding it continues to have, practically speaking, very nearly the same specific gravity as at the surface. Without this valuable property possessed by water, the hydraulic press would not exist.

The strange blunder here described was participated in by some of the most distinguished naval men. As an instance, even at quite a recent period, Captain Marryat, R.N., the famous nautical author, writes of the sea :—" What a mine of wealth must lie buried in its sands. What riches lie entangled amongst its rocks, or *remain suspended* in its unfathomable gulf, *where the compressed fluid is equal in gravity to that which it encircles.*" [1]

To obviate this non-existent difficulty, it was gravely proposed to festoon the cable across, at a given maximum depth, between buoys, and floats, or even parachutes—at which ships might call, hook on, and telegraphically talk to shore!

Others, again, proposed to apply gummed cotton

[1] *The Pirate*, by Capt. Marryat, p. 2.

to the outside of the cable in connection with the above buoying system. The idea was that the gum (or glue) would gradually dissolve, and so let the cable down "quietly"!

One naval officer of eminence urged the employment of an immense floating cylinder, on which the cable was to be wound. This cylinder was then to be towed across the Atlantic, unrolling the cable in its progress. The unwieldiness of such a cylinder—with some 2,500 tons of cable encircling it, in addition to its own weight, and the practical impossibility of regulating delivery in its revolutions, or of dealing with it even in an ordinary roughish sea, did not appear to have been held of much account by the projector.

Then, again, the claims of light cables were brought forward, cables that could not even be made to sink at a suitable angle to dip into the irregularities of the bottom—let alone the fact that their strength, notwithstanding their low specific gravity, was altogether insufficient.

It was also suggested that the proper place to pay out was from the centre of the ship, as the point of least motion, and therefore least liable to damage the cable; and it was proposed to have an opening in the middle to let it down. But as the cable in paying out leaves the ship at an angle only a little removed from the horizontal, the absurdity of such a proposition is manifest.

Again, a trail, or flexible pipe, was strongly

advocated, "to hang down from the ship's stern
to the bottom of the sea, through which the cable
was to be allowed to pass." But the promoters
of this plan omitted to consider the effects of the
friction resulting from 2,000 miles of cable passing through it. Of whatever substance such a
trail might be made, a day or two's rubbing of
the cable would have worn it through.

As an instance of the wild notions prevailing
in the mind of one gentleman with a proposed
invention, to whom was shown an inch specimen
of the cable, he remarked to Charles Bright:
"Now I understand how you stow it away on
board. You cut it up into bits beforehand, and
then join up the pieces as you lay."

Some again absolutely went so far as to take
out patents for converting the laying vessel into
a huge factory, with a view to making the cable
on board in one continuous length, and submerging it during the process !!

Before dismissing the subject of Bright's worries
in this direction, a very interesting case may
be mentioned. It is probably unparalleled for
lengthy perseverance in attempting to raise capital
speedily.

The person in question had opened his campaign
in May, 1857, with a lengthy statement of nearly
2,000 words—first with reference to his previous
career, and then describing an invention he had
made "by which alone the cable could be laid."

Next followed an assertion that no soundings could have been obtained across the Atlantic as "both modern science and actual experiment demonstrate that, long before any such depths could be reached, the lead must necessarily have displaced its own specific gravity in so dense a medium as water, and consequently at once then stop, remaining suspended equilibrated."

After the above introductory remarks, this gentleman asked for "a substantial sum down" for his "invention." He refused, however, to reveal its nature.

Charles Bright replied: "It is impossible to form any idea of the advantage, or otherwise, of the scheme you allude to, without some information respecting it; and I would therefore suggest that your best course would be to apply for provisional protection for your invention. You could then with perfect safety to yourself, fully unfold your views."

"I cannot agree with your judgment respecting the accuracy of the soundings reported in the Atlantic Ocean by Lieutenant Maury, Captain Berryman, and other officers. I have, indeed, a somewhat convincing proof of the view I take in my possession, viz., specimens of the bottom taken up at each of the soundings made by the *Arctic*, which I shall be most happy to show you."

This was only the preliminary skirmish so far as the man of war was concerned. There was immediately another communication from him of over a thousand words, mostly occupied in repeating that soundings could not possibly be taken, and in decrying Lieutenant Maury and Captain Berryman. He finished by saying that he had no intention of delaying "a great cosmographical benefit by taking out a patent." He then reiterated his proposal for a large sum down.

Bright could only answer : " I hope you will see the impossibility of the Company entering into any arrangement for the use of a scheme which they know nothing whatever about."

Bright then tried to close the correspondence ; but it was of no avail, for his note only produced the same day a fourth still longer letter still sticking to his main point—"money down, invention on trust." He also alleged that "the Atlantic Cable, from its trivial specific gravity, can never, no matter how weighted, sink more than about 300 fathoms."

Finding that nobody paid much attention to him. this gentleman then proceeded to put his statements into a book of some 200 pages, the publication of which, however, was stopped on the score of being libellous.

Let us return to the active and practical preparations for the forthcoming expedition. It is difficult for the uninitiated to realise what these meant. They would have driven many crazy, if only on account of their vast and varied character. In this connection, Charles Bright notes in his diary :—

> It was only by dint of bribing, bullying, cajoling, and going day by day to see the state of things ordered, that anything is ready in time for starting.

He then says :—

> At first one goes nearly mad with vexation at the delays ; but soon one finds that they are the rule, and then it becomes necessary to feign a rage one does not feel.

Further :—

> I look upon it as the natural order of things that if I give an order it will not be carried out ; or, if by accident it is carried out, it will be carried out wrongly. The only remedy is to watch the performance at every stage.

All this incessant toil seems to have additionally inspired the following note :—

> When idle, one can love, one can be good, feel kindly to all, devote oneself to others, be thankful for existence, educate one's mind, one's heart. one's body.[1] When busy, one sometimes seems too busy to indulge in any of these pleasures.

---

[1] The truth of this need not necessarily conflict with the fact that, in most instances, the *permanently* idle find no time for any of the above virtues.—AUTHORS.

THE ATLANTIC CABLE

As soon as one of the machines for paying out the cable was completed and set up in working order, the following letter was written :—

April 10*th.*

To the Managing Committee of the Atlantic Telegraph Company.

GENTLEMEN,—The first of the two paying-out machines in course of construction by Messrs. Easton & Amos will be finished to-night, and we shall probably have it running on Monday ; but I should wish to have three clear days at least for experimental purposes before any persons, except the directors and officials of the Company, are admitted to the works, after that it will be ready for the inspection of such visitors as you may wish should see it ; but it must be taken down with the least possible delay to allow of the other machine being put together in its place.

The *Adonis* [1] will commence loading the cable at Greenwich on Monday week, and as the machine will have to be taken to pieces and shipped on board her (which must be done very carefully, lest a fracture of any of the large castings should throw us back), there will be very little time to spare for the visitors—probably not more than a day, or two days at the outside.

I think it very desirable that all who are about to be actively engaged in the work of paying out the cable, should see the machine running next week ; also that in the event of any suggestion emanating from them as to

---

[1] This vessel was to take the new length of additional cable round to the ships at Keyham, where the rest of the line had been stored since the previous expedition.

any improvement of the details, it may be considered before it is too late to make any additions to, or alterations in, the machines.

Our staff arrangements on board the two ships, as they stand at present, will be as follows : On board the *Agamemnon* I shall take watch and watch with Mr Canning, in charge of the entire ship as regards the paying out of the cable. Mr. Hoare and Mr. Moore[1] will be alternately charged, under the direction of whichever of us is on duty at the time, with the care and adjustment of the machinery. Mr. Clifford, who was with us throughout the operations of last year, and has been at Messrs. Easton & Amos' works during the experiments and construction of the new machinery, will also be on board the *Agamemnon*; the experience which he has acquired in all that relates to coiling and paying out the cable, and his knowledge of the details of the machines and engines, qualify him for rendering valuable assistance in the ensuing expedition in superintending the paying-out in the event of accident to Mr. Canning or myself, or of over fatigue, as well as in the management of machinery.

On board the *Niagara*, Mr. Everett and Mr. Woodhouse will have Captain Kell as an extra assistant. Mr. Follansbee, the chief engineer of the *Niagara*, and one of the additional assistant engineers will have the care of the brakes on that vessel.

The above arrangements will be very complete. I had thought at one time of advising that three separate watches should have followed each other on board each

---

[1] Engineers of H.M. Navy, commissioned to the *Agamemnon*, but just recently transferred to Charles Bright's staff for the coming expedition.

ship; but on reflection, I am of opinion that it will be better to do with as few men as the watch can be properly done by, with short watches to avoid any overstraining of our physical energies.

I would therefore suggest that Messrs. Canning and Woodhouse, as well as the two naval engineers from each ship, should be requested to come up on some day during the ensuing week to make themselves acquainted with the machinery, Captain Kell taking charge of the coiling on board in their absence. I am sure he will work doubly hard to prevent anything going wrong while they are away; and as he is more of a sailor than an engineer, I do not think it important that he should come up; but, if you consider it desirable, he could do so on their return.

There will be a good deal of expense about this, but I consider the outlay will be amply repaid by making the staff well acquainted with the machinery before the experimental trip.

Eight hundred and sixty miles were coiled on board the *Agamemnon* to noon yesterday, and 537 on board the *Niagara*.

I am, Gentlemen, yours faithfully,

CHARLES T. BRIGHT.

A few days later all the staff named in this letter inspected the working of the machine, whilst at the same time receiving instructions as above for the coming expedition, including the trial mentioned, during which a complete rehearsal was to be gone through of the various operations to be performed with the apparatus provided for the purpose.

No material alterations were suggested to the paying-out gear after it had once been seen up; and to the hauling-in gear of 1857, the alterations were mainly confined to placing the engines on deck instead of below.

On the other hand, Bright's arrangement for stowing the cable aboard on this occasion formed a subject for discussion at the hands of some of the naval officers concerned with the undertaking. Our young engineer had determined this time that the large coil in the hold of the *Agamemnon* must be made as truly circular and also as large as he had insisted on for the *Niagara* in the previous year. He also decided that a cone in the middle of each coil, and a large margin of space to the hatchway-eye above were both essential provisions for safe paying out. These alterations were all duly made, although one of the naval experts had expressed himself that the cable "should be stowed in long Flemish flakes." The same officer also considered that "no other machinery for paying out was necessary or desirable than a handspike to stop the egress of the cable." (!) Charles Bright, whilst always ready to listen to suggestions, had sometimes to remind his critics that "criticism is always easier than art." In order to support his convictions, and to refute those of others, he had constantly to apply to such famous engineers as Sir John Rennie, F.R.S., Mr. I. K. Brunel, F.R.S., and Mr. Peter Barlow, F.R.S., besides Messrs. Glass & Elliot, and Messrs. Newall.

Whilst the cable was stored in the tanks at Keyham Dockyard Mr. Whitehouse—partly in conjunction with Prof Thomson—took the opportunity of conducting a fresh series of experiments through the entire length with various apparatus and under various conditions. These experiments were more especially in the direction of testing, and improving on, the rate of working. As a result a speed of four words per minute were attained through the 2000 odd miles.

Since the manufacture of the cable in 1857, Prof. Thomson had become impressed with the conviction that the electric conductivity of copper varied greatly with its degree of purity. As a result of the professor's further investigations, the extra length of cable made for the coming expedition was subjected to systematic and searching tests for the purity and conductivity of the copper. Every hank of wire was tested; and all whose conducting power fell below a certain value rejected. Here then we have the first instance of an organised system of testing for conductivity at the cable factory—a system which has ever since been rigorously insisted on.

And now, in the spring of 1858, an invention was perfected that was destined to have a remarkable effect on submarine cable enterprise. For within about a year of his entering the ranks of telegraphic scientists, Professor Thomson (now Lord Kelvin) devised and perfected the mirror speaking instrument, then often described as the marine galvanometer,[1] of which it may fairly be said that it entirely revolutionised long distance signalling and electrical testing aboard ship.

The following description[2] may be interesting to some of our readers :—

"This most ingenious apparatus consists of a small and exceedingly light steel magnet with a tiny reflector or mirror fixed to it, both together weighing but a *single grain*, or thereabouts. This delicate magnet is suspended from its centre by a filament of silk, and surrounded by a coil of the thinnest copper wire, silk covered.

"When electricity passes through this surrounding coil of wire, the magnet and mirror take up a position of equilibrium between the elastic force of the silk and the deflecting force of the current from the cable circulating

---

[1] In those days all such instruments were spoken of as galvanometers, no matter for what purpose they were employed. Moreover, this instrument was also used sometimes for testing purposes. That which goes by the name of the Marine Galvanometer in the present day was not invented by Lord Kelvin till some years later, and was first used in practice at sea on the Persian Gulf cable expedition of 1864.

[2] *The Electric Telegraph*, by Edward B. Bright, M. Inst. C.E. (London : Crosby Lockwood & Son).

through the coil. A very weak current is sufficient to produce a slight, though nearly imperceptible, movement of the suspended magnet.

"A fine ray of light from a shaded lamp, behind a screen at a distance, is directed through a slot in the screen, thence to the open centre of the coil upon the mirror. It is then reflected back to a graduated scale upon the *dark* side of the lamp screen turned towards the coil.

"An exceedingly slight angle of motion on the part of the magnet is thus made to magnify the movement of the spot of light upon the scale, and to render it so considerable as to be readily noted by the eye of the operating clerk. The ray is brought to a focus by passing through a lens. By combinations of these movements of the speck of light (in length and duration) upon the index, an alphabet is readily formed.

"The magnet is artificially brought back to zero with great precision after each signal by the use of an adjustable controlling magnet.

"In a word, Professor Thomson's combined mirror telegraph and marine galvanometer transmits messages by multiplying and magnifying the signals through a cable, by the agency of imponderable light.

"The arrangement of the apparatus is shown on the following page :—

"*C* represents the galvanometer mounted upon a stand, opposite the screen *b*, and its graduated scale *a* at a short distance. A lamp is placed at the back of the scale : and immediately below the zero mark of the scale a small slot *g* is cut to allow a streak of light to pass from the lamp to the small suspended mirror in the galvanometer coil : *d* is a thumbscrew to regulate the position of the mirror magnet, and *e* is a small adjusting magnet outside the coil.

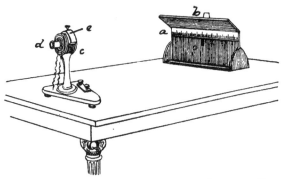

THE REFLECTING GALVANOMETER AND "SPEAKER."

"The next sketch shows the galvanometer coil *b* with its suspended magnet and mirror *a*.

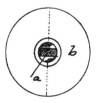

"Again, the diagram below serves to illustrate the manner in which the slightest movement of the mirror *A* makes the ray of light reflected from it traverse the graduated scale *B* at a distance."

It is only to be regretted that the electrician responsible for the subsequent working-through operations did not sooner appreciate the great beauties of the above apparatus, and the advantage of a small generating force such as it alone required.

The Board having conferred with the commanding officers of the telegraphic squadron, decided "that it would be desirable to begin paying out the cable in mid-ocean." Thus they reversed the starting from shore of the previous expedition. The latter, it will be remembered, was a concession to the electricians, though strongly opposed by Charles Bright and the whole of his engineering staff at the time, their reasons being (as aforesaid) the ability to choose the day for joining the ends in good weather ; the reduction of the time of the laying operation by one half, with thus a better chance of fine weather being maintained throughout the expedition ; and, finally, that the most difficult part of the work would be dealt with at first in the deepest water.

It was also arranged by Charles Bright that the main cable should be buoyed at each end, and the connection to it by the heavy cable from shore effected at the earliest opportunity afterwards.

Further progress is best reported, and an idea of the programme for future operations best set forth, in the following letter :—

SIR CHARLES TILSTON BRIGHT

May 5th.

To the Directors of the Atlantic Telegraph Company.

GENTLEMEN,—Since my last report to the Board of Directors the 300 miles of additional cable ordered from Messrs. Glass, Elliot & Co., have been finished.

The *Adonis* having completed her delivery at Keyham of the 400 miles previously made, proceeded to Greenwich on the 26th inst., and commenced coiling in the new length, which operation will be completed to-day.

One set of the new paying-out machinery having been constructed by Messrs. Easton & Amos, it was erected in connection with one of the old hauling-in machines, so that it might be tried as effectually as possible on shore prior to fitting it on board the ships. The result of a week's experiments, with various speeds and strains upon the cable by which it was driven, was most satisfactory. It has since been taken down, and is now being placed on a lighter for shipment on board the *Adonis*.

As our arrangements now stand, we should be ready to start on the experimental trip by the 25th of the present month; but the recently ordered length of 100 miles will detain us for a few days longer.

We shall reach water of 2,000 fathoms in depth within three days after leaving Plymouth.

We shall have about thirty-six miles of defective cable divided between the two ships for experimental purposes.

The longer the handling of this can be prolonged, without the cable parting the better; but I am not very sanguine about recovering it again after it has been paid out in water of such a depth, nor are any of the engineers who are co-operating with me in the work.

THE ATLANTIC CABLE

The chief points which it is desirable to rehearse are, the junction of the two cables, lowering the splice, paying out with the new machine, changing from one coil to another, stopping and allowing the cable to become vertical, buoying the cable, and hauling in again. All these, but the buoying and hauling-in were successfully accomplished last year; but it is important that the working of our new appliances should be tested at sea. It is to be regretted that the splice must be made in fine weather, and that our operations following upon it will therefore necessarily be carried out under the same circumstance; for I should have liked to try the paying out part of the trip under the worst possible conditions.

If we do not recover the cable, the experiments will not occupy over two days; and not so much if the cable is passed successfully from one ship to the other, and the splice made at the first attempt. But if we should meet with success in hauling in, we can repeat the junction, paying out, and buoying until we lose all the trial cable.

We cannot determine the time when the trial expedition will have returned, as the weather may not be suitable for making the splice when we get into deep water; but the shortest period will be seven days from the time of starting.

Should everything proceed to our satisfaction in the trial trip—and the weather promise so favourably that the captains of the ships composing the squadron consider it a suitable time for laying the cable—it appears to me very desirable that we should at once proceed with the work without returning to England, unless it is necessary for some alteration to be made in our arrangements.

This need not be publicly put forth, lest our return should imply a reverse. But if our experiments should be extended, by heavy weather preventing the splice being made, and we should find ourselves approaching the middle of June, it seems to me important that we should commence to lay the cable forthwith if recommended so to do by our nautical advisers.

1,260 miles of cable are coiled on board the *Agamemnon*, and 1,146 on board the *Niagara*.

I am, Gentlemen, yours faithfully,

CHARLES T. BRIGHT,

Engineer.

SECTION VII

The Trial Trip

All the 3,000 miles of cable was coiled into

DECK OF H.M.S. *AGAMEMNON*, SHOWING THE PAYING-OUT APPARATUS
(From the *Illustrated London News*)

the two large ships and the improved machinery fitted on board of them by the end of May. The paying-out apparatus mounted on the deck of the *Agamemnon* may be seen in the accompanying illustration, whilst we also show a view in section of the fore tanks of the *Niagara* (Captain W. L. Hudson, U.S.N.) when loaded with her cargo of cable.

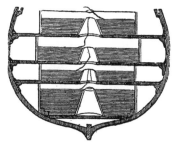

STOWAGE OF THE COILS IN THE FOREPART OF THE *NIAGARA*

The *Agamemnon* was this time in naval command of Captain (now Vice-Admiral) G. W. Preedy,[1] but her navigating master was Mr. H. A. Moriarty, R.N., as before. The accompanying sketch shows her loading.

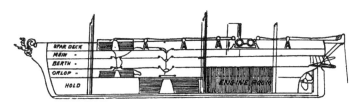

THE LOADING OF THE *AGAMEMNON*

[1] Captain Noddall, R.N., had in the meanwhile been appointed to another ship.

SIR CHARLES TILSTON BRIGHT

The rest of the telegraph squadron was made up of H.M.S. *Gorgon* (Captain Dayman, who had taken the soundings in the *Cyclops* the year before); H.M. paddle steamer *Valorous* (Captain W. C. Aldham), in place of U.S. frigate *Susquehanna*, and H.M. surveying steamer *Porcupine* (Captain H. C. Otter).

Thus equipped, the fleet again set forth from Plymouth on May 29th, 1858, but this time without any show of public enthusiasm. Charles Bright was accompanied by the staff, engineering and electrical, already referred to. With him on the *Agamemnon* were Mr. Samuel Canning (his chief assistant), and Mr. Henry Clifford; whilst, on the *Niagara* he was jointly represented by Messrs. W. E. Everett[1] and Mr. Henry Woodhouse, the former taking charge of the machinery, and the latter—with a greater experience in such work —of the cable. They were assisted by Captain John Kell. Mr. Cyrus Field also accompanied the *Niagara*, Mr. Whitehouse being again unable to take passage. Professor Thomson had this time consented to take a more active part in the electrical work. He, in fact, agreed to supervise the testing-room arrangements in the *Agamemnon.*[2]

---

[1] It had been felt that on an American war-ship, an American engineer officer should have some control, if possible.

[2] In adopting this course, the Professor showed what

whilst Mr. De Sauty and Mr. Laws—together with Mr. John Murray, of "Loop-test" fame—had the electrical force of the *Niagara* under their charge. The various gentlemen who were about to undertake the working of the cable were the same as those that set out on the previous expedition. The Press were again *en evidence* as regards both ships, the newspapers being represented by the same gentlemen as before.

Although the improved paying-out gear had passed through most satisfactory experiments at Messrs. Easton & Amos' works, it was arranged by Charles Bright, as already stated, to test it practically in very deep water—besides making splices at sea, picking up, buoying and exercising all hands in their work generally—before commencing to lay in mid-Atlantic.

So the cable-laden ships, with H.M.S. *Valorous* and H.M.S. *Gorgon* as consorts, first made a trial trip to the Bay of Biscay as far as lat. 47° 12′ N., long. 9° 32′ W., about 120 miles north-west of

manner of man he was, in taking up, at the solicitation of his co-directors, Dr. Whitehouse's duties. Professor Thomson undertook the above arduous position without recompense of any kind, though involving the temporary abandonment of his academical work and also his scientific researches.

Corunna, where the *Gorgon* got soundings of 2,530 fathoms, or nearly three statute miles, in depth.

The *Agamemnon* and *Niagara* were then backed close together, stern on, and a strong hawser was passed between them.

Each ship had on board some defective cable for the experiment about to be conducted, the nature

THE TRIAL TRIP.

of which had been stated in Charles Bright's last letter to his directors. The further proceedings may now be observed from a perusal of Bright's diary, written aboard the *Agamemnon*.

Monday, May 31st.—10 a.m., hove to, lat. 47°11', long. 9°37'. Up to mid-day engaged in making splice between experimental cable in fore coil and that in main hold, besides other minor operations. In afternoon, getting hawser from *Niagara*, and her portion of cable to make, joint, and splice. 4 p.m., commenced splice; 5.15, splice completed; 5.25, let go splice frame (weight 3 cwt.) over gangway, amidships, starboard side.[1] 5.30, after getting splice frame (containing the splice), clear of the ship and lowering it to the bottom, each vessel (then about a quarter of a mile apart) commenced paying out in opposite directions.

9 p.m., got on board *Niagara's* warp and her end of cable, to make another splice for 2nd experiment.

June 1st.—1 a.m. (night), electrical continuity gone, the cable having parted after two miles in all had been paid out.[2]

Since 1 a.m., engaged in hauling in our cable. Recovered all our portion, and even managed to heave up the splice frame (in perfect condition), besides 100 fathoms of *Niagara's* cable, which she had parted. Fastened splice to stern of vessel and ceased operations.

9.23 a.m., 2nd experiment. Started paying out again. Weather very misty.

9.40, one mile paid out at strain 16 cwt.; angle of cable 16° with the horizon; running out straight; rate of ship 2, cable 3.

9.45, changed to lower hold. 9.56, two miles out; last mile in 16½ minutes; strain 17 to 20 cwt.; angle of cable 20°. 10.10, last three miles out in 14 minutes.

---

[1] The character of this splice-frame, and the object for which it was provided, has already been explained in connection with the preparations made in 1857 (see p. 156).

[2] This of course did not in any way come as a surprise, for the length of cable employed for these experiments had long since been condemned as imperfect.

10.32 a.m., four and half miles out. 3rd experiment stopped ship, lowered guard, stoppered cable.

10.50, buoy let go, strain 16 cwt. when let go, the cable being nearly up and down. 11.6, running at rate of 5⅓ knots paying out, strain 21 to 23 cwt, varying.[1] Cable shortly afterwards parted, through getting jammed in the machinery.

The subsequent experiments were mainly in the direction of buoying, picking up, and passing the cable from the stern to the bow sheave for picking up, All of these operations were in turn successfully performed; and, finally, in paying out, a speed of seven knots was attained without difficulty."

And now, the programme being exhausted, there was nothing left to be done but to return to Plymouth.

On the whole the trip had proved eminently satisfactory. The paying out machinery had worked well, the various engineering operations had been successfully performed, and the electrical working through the whole cable was perfect. Professor Thomson had brought with him that offspring of his brain—his reflecting and testing

---

[1] During all this time electrical communication had been maintained between the ships; and it is somewhat remarkable that through this more or less damaged cable, the electricians were able (as noted in Bright's Diary) to work a needle instrument and obtain a deflection of 70°.

instrument[1]—and it was this that had given such excellent results.

Section VIII

The Storm

The "wire ships" thus additionally experienced arrived at Plymouth on Thursday, June 3rd. The results were duly reported by Charles Bright, and some further arrangements were made, principally connected with the electrical department.

A week later, *i.e.* on Thursday, June 10th, having taken in a fresh supply of coal, the expedition again left England "with fair skies and bright prospects." The barometer standing at 33·64, it was an auspicious start in what was declared by a consensus of nautical authorities to be the best time of the year for the Atlantic.

This prognostication was doomed to a terrible disappointment, for the voyage nearly ended in the *Agamemnon* "turning turtle." She was repeatedly almost on her beam ends, the cable was partly shifted,[2]

[1] This apparatus for the first time turned to practical account the general principle of Gauss & Weber's heavily constructed, and comparatively sluggish, telegraph of 1837.

[2] The load of cable made all the difference when brought into comparison with an ordinary ship under stress of weather. It was bad enough to cruise with a dead weight forward of some 250 tons—a weight under which

and a large number of those on board were more or less seriously injured.[1]

Charles Bright, with Messrs. Canning and Clifford, were—as during the trial trip—on the *Agamemnon*, and also Professor Thomson, who again took charge of the electricians' department, Mr. Whitehouse being ashore. Messrs. Everett and Woodhouse were once more on the *Niagara*, with Mr. De Sauty superintending the signalling.

her planks gaped an inch apart, and her beams threatened daily to give way. But when to these evils were added the fear that in some of her heavy rolls the whole mass would slip and take the vessel's side out, it will be seen that this precious coil was justly regarded as a standing danger—the millstone about the necks of all on board. And so it is sometimes with telegraph ships—as regards the dead weight of cable—even in the present day, when compared with the risks run by ordinary sea-going vessels.

[1] Oddly enough, owing to the fact that the *Agamemnon* had scant accommodation left for fuel, every one at the start was bemoaning the entire absence of breeze. There were some even who, never having been at sea before, muttered rash hopes about meeting an Atlantic gale. Their wishes were soon to be completely realised.

Others there were who talked of the possibility of light winds, and of encountering some delay from calms and sultry weather ; but a gale—a regular Atlantic storm— the very idea was food for laughter ! It was to be a yachting cruise—a mere summer trip—and any talk of waterproofs and sou'-westers would not have been more out of place in a drawing-room than on the deck of the *Agamemnon*.

Mr. Cyrus Field, as before, sailed in the American ship.

In order that laying operations should be started by the two ships in mid-ocean, it was arranged that the entire fleet should meet in lat. 52° 2′, and long. 33° 18′ as a *rendezvous*. The *Porcupine*, the smallest ship of the squadron, had been sent to St. John's, Nova Scotia, with orders to meet the *Niagara* on her way to Trinity Bay. Besides the laying vessels, there were the *Valorous* and the *Gorgon*, the former acting as an escort to the *Agamemnon*, and the latter doing similar duty for the *Niagara*.

As it is impossible to follow the movements of more than one ship at a time, and as the vessel which Charles Bright sailed with—the *Agamemnon* —had the more exciting experience, we will confine our attention to her up to the date of the *rendezvous*.

The day after starting there was no wind, but on the Saturday, the 12th, a breeze sprung up, and, with screw hoisted and fires raked out, the *Agamemnon* bowled along at a rare pace under royals and studding sails. The barometer fell fast, and squally weather coming on with the boisterous premonitory symptoms of an Atlantic gale, even those least versed in such matters could see at a glance that they were "in for it."

On Sunday the sky was a wretched mist—half rain, half vapour—through which the attendant vessels loomed faintly like shadows, apparently as unsubstantial as the *Flying Dutchman.* The gale increased; till at four in the afternoon the good ship was rushing through the foam under close-reefed topsails and foresail.

That night the storm got worse, and most of the squadron gradually parted company. The ocean resembled one vast snowdrift, the whitish glare from which—reflected from the dark clouds that almost rested on the sea—had a tremendous and unnatural effect, as if the ordinary laws of nature had been reversed.

Very heavy weather continued till the following Sunday, the 20th June, which ushered in as fierce a storm as ever swept over the Atlantic; and the narrative of this fight of nautical science with the elements may best be continued in the words of Mr. Nicholas Woods, who, representing and reporting for *The Times,* was an eye-witness throughout, especially as it is probably the most intensely realistic description of a storm that has ever been written :—

The *Niagara,* which had hitherto kept close—whilst the other smaller vessels had dropped out of sight—began to give us a very wide berth and, as darkness increased, it was a case of every one for themselves.

Our ship, the *Agamemnon,* rolling many degrees—not every one can imagine how she went at it that night— was labouring so heavily that she looked like breaking up.

The massive beams under her upper deck coil cracked

and snapped with a noise resembling that of small artillery, almost drowning the hideous roar of the wind as it moaned and howled through the rigging; jerking and straining the little storm-sails as though it meant to tear them from the yards. Those in the improvised cabins on the main deck had little sleep that night, for the upper deck planks above them were "working themselves free," as sailors say; and, beyond a doubt, they were infinitely more free than easy, for they groaned under the pressure of the coil with a dreadful uproar, and availed themselves of the opportunity to let in a little light, with a good deal of water, at every roll. The sea, too, kept striking with dull heavy violence against the vessel's bows, forcing its way through hawse-holes and ill-closed ports with a heavy slush; and thence, hissing and winding aft, it roused the occupants of the cabins aforesaid to a knowledge that their floors were under water, and that the flotsam and jetsam noises they heard beneath were only caused by their outfit for the voyage taking a cruise of its own in some five or six inches of dirty bilge. Such was Sunday night, and such was a fair average of all the nights throughout the week, varying only from bad to worse. On Monday things became desperate.

The barometer was lower and, as a matter of course, the wind and sea were infinitely higher than the day before. It was singular, but at 12 o'clock the sun pierced through the pall of clouds, and shone brilliantly for half an hour, and during that brief time it blew as it has not often blown before. So fierce was this gust, that its roar drowned every other sound, and it was almost impossible to give the watch the necessary orders for taking in the close-reefed foresail; which, when furled, almost left the *Agamemnon* under bare poles, though still surging through the water at speed. This gust passed, and the usual gale set in—now blowing steadily from the south-west, and taking us more and more out of our course each minute. Every hour the storm got worse, till towards five in the afternoon, when it seemed at its height, and raged with such a violence of wind and sea, that matters really looked "desperate" even for such a strong and large ship as the *Agamemnon*. The upper deck coil had strained her decks throughout excessively; and, though this mass, in theory, was supposed to prevent her rolling so quickly and heavily as she would have done without it, yet still she heeled over to such an alarming extent that fears of the coil itself shifting again occupied every mind, and it was accordingly strengthened with additional shores bolted down to the deck. The space occupied by the main coil below had deprived the *Agamemnon* of several of her coal bunkers, and in order to make up for this deficiency, as well as to endeavour to counterbalance the immense mass which weighed her down by the head, a large quantity of coals had been stowed on the deck aft. On each side of her main deck were thirty-five tons, secured in a mass, while on the lower deck ninety tons were stowed away in the same manner. The precautions taken to secure these huge masses also required attention as the great ship surged from side to side. But these coals seemed secure; and were so, in fact, unless the vessel should almost capsize—an unpleasant alternative which no one certainly anticipated then. Everything, therefore, was made "snug," as sailors call it; though their efforts by no means resulted in the comfort which might have been expected from the term. The night, however, passed over without any mischance beyond the smashing of all things incautiously left loose and capable of rolling, and one or two attempts which the *Agamemnon* made in the middle watch to turn bottom upwards. In all other matters it was the mere ditto of Sunday night; except, perhaps, a little worse, and certainly much more wet below. Tuesday the gale continued with almost unabated force; though the barometer had risen to 29·30, and there was sufficient sun to take a clear observation, which showed our distance from the rendezvous to be 563 miles. During this afternoon the *Niagara* joined company, and, the wind going more ahead, the *Agamemnon* took to violent pitching, plunging steadily into the trough of the sea as if she meant to break her back and lay the Atlantic cable in a heap. This change in her motion strained and taxed every inch of timber near the coils to the very utmost. It was curious to see how they worked and bent as the *Agamemnon* went at everything she met head first. One time she pitched so heavily as to break one of the main-beams of the lower deck, which had to be shored with screw-jacks forthwith. Saturday, the 19th of June, things looked a little better. The barometer seemed inclined to go up and the sea to go down; and for the first time that morning—since the gale began some six days previous—the decks could be walked with tolerable comfort and security. But, alas! appearances are as deceitful in the Atlantic as elsewhere; and during a comparative calm that afternoon the glass fell lower, while a thin line of black haze to windward seemed to grow up into the sky, until it covered the heavens with a sombre darkness, and warned us that, after all, the worst was yet to come. There was much heavy rain that evening, and then the wind began—not violently, nor in gusts, but with a steadily increasing force, as if the gale was determined to do its work slowly but do it well. The sea was " ready-built to hand," as sailors say, so at first the storm did little more than urge on the ponderous masses of water with redoubled force, and fill the air with the foam and spray it tore from their rugged crests. By-and-by, however, it grew more dangerous, and Captain Preedy himself remained on deck throughout the middle watch, for the wind was hourly getting worse and worse, and the *Agamemnon*, rolling thirty degrees each way, was straining to a dangerous extent.

At 4 a.m. sail was shortened to close-reefed fore and maintopsails and reefed foresail—a long and tedious job, for the wind so roared and howled, and the hiss of the boiling sea was so deafening, that words of command were useless, and the men aloft, holding on with all their might to the yards as the ship rolled over and over almost to the water, were quite incapable of struggling with the masses of wet canvas, that flapped and plunged as if men and yards and everything were going away together. The ship was almost as wet inside as out—and so things wore on till 8 or 9 o'clock, everything getting adrift and being smashed, and every one on board jamming themselves up in corners or holding on to beams to prevent their going adrift likewise. At 10 o'clock the *Agamemnon* was rolling and labouring fearfully, with the sky getting darker, and both wind and sea increasing every minute. At about half-past 10 o'clock three or four gigantic waves were seen approaching the ship, coming slowly on through the mist nearer and nearer, rolling on like hills of green water, with a crown of foam that seemed to double their height. The *Agamemnon* rose heavily to the first, and then went down quickly into the deep trough of the sea, falling over as she did so, so as almost to capsize completely on the port side. There was a fearful crashing as she lay over this way, for everything broke adrift, whether secured or not, and the uproar and confusion were terrific for a minute; then back she came again on the starboard beam in the same manner, only quicker, and still deeper

H.M.S. *AGANEMNON* IN A STORM.

247

than before. Again there was the same noise and crashing, and the officers in the ward-room, who knew the danger of the ship, struggled to their feet and opened the door leading to the main deck. Here, for an instant, the scene almost defies description. Amid loud shouts and efforts to save themselves, a confused mass of sailors, boys, and marines, with deck-buckets, ropes, ladders, and everything that could get loose, and which had fallen back again to the port side, were being hurled again in a mass across the ship to starboard. Dimly, and only for an instant, could this be seen, with groups of men clinging to the beams with all their might, with a mass of water, which had forced its way in through ports and decks, surging about; and then, with a tremendous crash, as the ship fell still deeper over, the coals stowed on the main deck broke loose, and, smashing everything before them, went over among the rest to leeward. The coal-dust hid everything on the main deck in an instant; but the crashing could still be heard going on in all directions, as the lumps and sacks of coal, with stanchions, ladders, and mess-tins, went leaping about the decks, pouring down the hatchways, and crashing through the glass sky-lights into the engine-room below. Still it was not done, and, surging again over another tremendous wave, the *Agamemnon* dropped down still more to port, and the coals on the starboard side of the lower deck gave way also, and carried everything before them. Matters now became most serious, for it was evident that two or three more such lurches and the masts would go like reeds, while half the crew might be maimed or killed' below. Captain Preedy was already on the poop, with Lieutenant Gibson, and it was " Hands, wear ship," at once; while Mr. Brown, the indefatigable chief engineer, was ordered to get up steam immediately. The crew gained the deck with difficulty, and not till after a lapse of some minutes; for all the ladders had been broken away, the men were grimed with coal-dust, and many bore still more serious marks upon their faces of how they had been knocked about below. There was some confusion at first, for the storm was fearful. The officers were quite inaudible; and a wild dangerous sea, running mountains

THE *AGAMEMNON* STORM : COALS ADRIFT.

high, heeled the great ship backwards and forwards, so that the crew were unable to keep their feet for an instant, and in some cases were thrown across the decks in a fearful manner. Two marines went with a rush head-foremost into the paying-out machine, as if they meant to butt it over the side; yet, strange to say, neither the men nor the machine suffered. What made matters worse, the ship's barge, though lashed down to the deck, had partly broken loose ; and dropping from side to side as the vessel lurched, it threatened to crush any who ventured to pass it. The regular discipline of the ship, however, soon prevailed, and the crew set to work to wear round the ship on the starboard tack, while Lieutenants Robinson and Murray went below to see after those who had been hurt, and about the number of whom extravagant rumours prevailed among the men. There were, however, unfortunately but too many. The marine sentry outside the ward-room door on the main deck had not had time to escape, and was completely buried under the coals. Some time elapsed before he could be got out, for one of the beams used to shore up the sacks, which had crushed his arm very badly, still lay across the mangled limb, jamming it in such a manner that it was found impossible to remove it without risking the man's life. Saws therefore had to be sent for, and the timber sawn away before the poor fellow could be extricated. Another marine on the lower deck endeavoured to save himself by catching hold of what seemed a ledge in the planks ; but, unfortunately, it was only caused by the beams straining apart, and, of course, as the *Agamemnon* righted they closed again, and crushed his fingers flat. One of the assistant-engineers was also buried among the coals on the lower deck, and sustained some severe internal injuries. *The lurch of the ship was calculated at forty-five degrees each way for five times in rapid succession.* The galley coppers were only half filled with soup ; nevertheless, it nearly all poured out, and scalded some of the poor fellows who were extended on the decks, holding on to anything in reach. These, with a dislocation, were the chief casualties ; but there were others of bruises and contusions, more or less severe, and, of course, a long list of escapes more marvellous than any injury. One poor fellow went head-first from the main deck into the hold without being hurt; and one on the orlop deck was "chevied" about for some ten minutes by three large casks of oil which had got adrift, and any one of which would have flattened him like a pancake had it overtaken him.

As soon as the *Agamemnon* had gone round on the other tack the *Niagara* wore also, and bore down as if to render assistance. She had witnessed our danger, and, as we afterwards learnt, imagined that the upper deck coil had broken loose and that we were sinking. Things, however, were not so bad as that, though they were bad enough, heaven knows, for everything seemed to go wrong that day. The upper deck coil had strained the ship to the very utmost, but still held on fast. But not so the coil in the main hold, which had begun to get adrift, and the top kept working and shifting over from side to side, as the ship lurched, until some forty or fifty miles were in a hopeless state of tangle, resembling nothing so much as a cargo of live eels; and there was every prospect of the tangle spreading deeper and deeper as the bad weather continued.

Going round upon the starboard tack had eased the ship to a certain extent, but with such a wind and such a sea, both of which were getting worse than better, it was impossible to effect much for the *Agamemnon's* relief, and so by twelve o'clock she was rolling almost as badly as ever. The crew, who had been at work since nearly four in the morning, were set to clear up the decks from the masses of coal that covered them; and while this was going forward a heavy sea struck the stern, and smashed the large iron guard-frame, which had been fixed there to prevent the cable fouling the screw in paying-out. Now that one side had broken, it was expected every moment that other parts would go, and the pieces hanging down either smash the screw or foul the rudderpost. It is not over-estimating the danger to say, that had the latter accident occurred in such a sea, and with a vessel so overladen, the chances would have been sadly against the *Agamemnon* ever appearing at the *rendezvous*. Fortunately it was found possible to secure the broken frame temporarily with hawsers so as to prevent its dropping further, though nothing could hinder the fractured end from striking against the vessel's side with such force as to lead to serious apprehensions that it would establish a dangerous leak under water. It was near three in the afternoon before this was quite secured, the gale still continuing, and the sea running even worse. The condition of the masts too at this time was a source of much anxiety both to Captain Preedy and Mr. Moriarty, the master. The heavy rolling had strained and slackened the wire shrouds to such an extent that they had become perfectly useless as supports. The lower masts bent visibly at every roll, and once or twice it seemed as if they must go by the board. Unfortunately nothing whatever could be done to relieve this strain by sending down any of the upper spars, since it was only her masts which prevented the ship rolling still more and quicker ; and so every one knew that if once they were carried away it might soon be all over with the ship, as then the deck coil could not help going after them. So there was nothing for it but to watch in anxious silence the way they bent and strained, and trust in Providence for the result. About six in the evening it was thought better to wear ship again and stand for the *rendezvous* under easy steam, and her head accordingly was put about and once more faced the storm. As she went round, she of course fell into the trough of the sea again ; and rolled so awfully

SIR CHARLES TILSTON BRIGHT

as to break her waste steam-pipe, filling her engine-room
with steam, and depriving her of the services of one boiler
when it was sorely needed. The sun set upon as wild and
wicked a night as ever taxed the courage and coolness
of a sailor. There were, of course, men on board who
were familiar with gales and storms in all parts of the
world; and there were some who had witnessed the
tremendous hurricane which swept the Black Sea on the
memorable 14th of November, when scores of vessels were
lost and seamen perished by the thousands. But of all
on board none had ever seen a fiercer or more dangerous
sea than raged throughout that night and the following
morning, tossing the *Agamemnon* from side to side like a
mere plaything among the waters. The night was thick
and very dark, the low black clouds almost hemming the
vessel in; now and then a fiercer blast than usual drove
the great masses slowly aside, and showed the moon, a
dim, greasy blotch upon the sky, with the ocean, white
as driven snow, boiling and seething like a cauldron. But
these were only glimpses, which were soon lost, and again
it was all darkness, through which the waves, suddenly
upheaving, rushed upon the ship as though they must
overwhelm it, and dealing it one staggering blow, went
hissing and surging past into the darkness again. The
grandeur of the scene was almost lost in its dangers
and terrors, for of all the many forms in which death
approaches man there is none so easy in fact, so terrific in
appearance, as death by shipwreck.

Sleeping was impossible that night on board the
Agamemnon. Even those in cots were thrown out, from
their striking against the vessel's side as she pitched.
The berths of wood fixed athwartships in the cabins on
the main deck had worked to pieces. Chairs and tables
were broken, chests of drawers capsized, and a little surf

was running over the floors of the cabins themselves, pouring miniature seas into portmanteaus, and breaking over carpet-bags of clean linen. Fast as it flowed off by the scuppers it came in faster by the hawse-holes and ports, while the beams and knees strained with a doleful noise, as though it was impossible they could hold together much longer; and on the whole it was as miserable and even anxious a night as ever was passed on board any line-of-battle ship in Her Majesty's service. Captain Preedy never left the poop all night; though it was hard work to remain there, even holding on to the poop-rail with both hands. Morning brought no change, save that the storm was as fierce as ever; and though the sea could not be higher or wilder, yet the additional amount of broken water made it still more dangerous to the ship. Very dimly, and only now and then through the thick scud, the *Niagara* could be seen—one moment on a monstrous hill of water and the next quite lost to view, as the *Agamemnon* went down between the waves. But even these glimpses showed us that our Transatlantic consort was plunging heavily, shipping seas, and evidently having a bad time of it; though she got through it better than the *Agamemnon*, as of course she could, having only the same load, though 2,000 tons larger. Suddenly it came on darker and thicker, and we lost sight of her in the thick spray, and had only ourselves to look after. This was quite enough, for every minute made matters worse, and the aspect of affairs began to excite most serious misgivings in the minds of those in charge. The *Agamemnon* is one of the finest line-of-battle ships in the whole navy; but in such a storm, and so heavily overladen, what could she do but make bad weather worse, and strain and labour and fall into the trough of the sea, as if she were going down head foremost? Three or four hours more and the vessel had borne all she could bear with safety. The masts were rapidly getting worse, the deck coil worked more and more with each tremendous plunge ; and, even if both these held, it was evident that the ship itself would soon strain to pieces if the weather continued so. The sea, forcing its way through ports and hawse-holes, had accumulated on the lower deck to such an extent that it flooded the stoke-hole, so that the men could scarcely remain at their posts. Everything went smashing and rolling about. One plunge put all the electrical instruments *hors de combat* at a blow, and staved some barrels of strong solution of sulphate of copper, which went cruising about, turning all it touched to a light pea-green. By-and-by she began to ship seas. Water came down the ventilators near the funnel into the engine-room. Then a tremendous sea struck her forward, drenching those on deck, and leaving them up to their knees in water, and the least versed on board could see that things were fast going to the bad unless a change took place in the weather or the condition of the ship. Of the first there seemed little chance. The weather certainly showed no disposition to clear—on the contrary, livid-looking black clouds seemed to be closing round the vessel faster and faster than ever. For the relief of the ship, three courses were open to Captain Preedy—one to wear round and try her on the starboard tack, as he had been compelled to do the day before ; another, to fairly run for it before the wind ; and, the third and last, to endeavour to lighten the vessel by getting some of the cable overboard. Of course the latter would not have been thought of till the first two had been tried and failed —in fact, not till it was evident that nothing else could save the ship. Against wearing round there was the danger of her again falling off into the trough of the sea, losing her masts, shifting the upper deck coil, and so finding her way to the bottom in ten minutes; while to attempt running before the storm with such a sea on was to risk her stern being stove in, and a hundred tons of water added to her burden with each wave that came up afterwards, till the poor *Agamemnon* went under them all for ever. A little after ten o'clock on Monday, the 21st, the aspect of affairs was so alarming that Captain Preedy resolved at all risks to try wearing the ship round on the other tack. It was hard enough to make the words of command audible, but to execute them seemed almost impossible. The ship's head went round enough to leave her broadside on to the seas, and then for a time it seemed as if nothing could be done. All the rolls which she had ever given on the previous day seemed mere trifles compared with her performances then. Of more than 200 men on deck at least 150 were thrown down, and falling over from side to side in heaps, while others, holding on to ropes, swung to and fro with every heave. It really appeared as if the last hour of the stout ship had come, and to this minute it seems almost miraculous that her masts held on. Each time she fell over her main chains went deep under water. The lower decks were flooded, and those above could hear by the fearful crashing —audible amid the hoarse roar of the storm—that the coals had got loose again below, and had broken into the engine-room, and were carrying all before them. During these rolls the main deck coil shifted over to such a degree as quite to envelope four men, who, sitting on the top, were trying to wedge it down with beams. One of them was so much jammed by the mass which came over him that he was seriously contused. He had to be removed to the sick bay, *making up the sick list to forty-five*, of which ten were from injuries caused by the rolling of the ship, and very many of the rest from continual fatigue and exposure during the gale. Once round on the starboard tack, and it was seen in an instant that the ship was in no degree relieved by the change. Another heavy sea struck her forward, sweeping clean over the fore part of the vessel, and carrying away the woodwork and platforms which had been placed there round the machinery for under-running. This and a few more plunges were quite sufficient to settle the matter, and at last, reluctantly, Captain Preedy succumbed to the storm he could neither conquer or contend against. Full steam was got on, and, with a foresail and foretopsail to lift her head, the *Agamemnon* ran before the storm, rolling and tumbling over the huge waves at a tremendous pace. It was well for all that the wind gave this much way on her, or her stern would infallibly have been stove in. As it was, a wave partly struck her on the starboard quarter, smashing the quarter galley and ward-room windows on that side ; and sending such a sea into the ward-room itself as literally almost to wash two officers off a sofa on which they were resting on that side of the ship. This was a kind of parting blow, for the glass began to rise, and the storm was evidently beginning to moderate ; and although the sea still ran as high as ever, there was less broken water, and altogether, towards midday, affairs assumed a better and more cheerful aspect. The ward-room that afternoon was a study for an artist ; with its windows half darkened and smashed, the sea water still slushing about in odd corners, with everything that was capable of being broken strewn over the floor in pieces, and some fifteen or twenty officers, seated amid the ruins, holding on to the deck or table with one hand, while with the other they contended at a disadvantage with a tough meal—the first which most had eaten for twenty-four hours. Little sleep had been indulged in, though much lolloping about. Those, however, who prepared themselves for a night's rest in their berths rather than at the ocean bottom, had great difficulty in finding their day garments of a morning. The boots especially went astray, and got so hopelessly mixed that the man who could "show up" with both pairs of his own was, indeed, a man to be congratulated.

But all things have an end, and this long gale—of over a week's duration—at last blew itself out, and the weary ocean rocked itself to rest.

Throughout the whole of Monday the *Agamemnon* ran before the wind, which moderated so much that at 4 a.m. on Tuesday her head was again put about ; and for the second time she commenced beating up for the *rendezvous*, then some 200 miles further from us than when the storm was at its height on Sunday morning. So little was gained against this wind, that Friday, the 25th—sixteen days after leaving Plymouth—still found us some fifty miles from the *rendezvous*. So it was determined to get up steam and run down on it at once. As we approached the place of meeting the angry sea went down. The *Valorous* hove in sight at noon; in the afternoon the *Niagara* came in from the north ; and at even, the *Gorgon* from the south; and then, almost for the first time since starting, the squadron was reunited near the spot where the great work was to have commenced fifteen days previously—as tranquil in the middle of the Atlantic as if in Plymouth Sound.

SECTION IX

The Renewed Effort

That evening the four vessels lay together side by side, and there was such a stillness in the sea and air as would have seemed remarkable even on an inland lake. On the Atlantic, and after what had been so lately experienced, it seemed almost unnatural.

The boats were out, and the officers were passing from ship to ship, telling their experiences of the voyage, and forming plans for the morrow. Captain Preedy had a sorry tale to tell. The strain to which the *Agamemnon* had been subjected during the storm—by the great weight, rendering her almost unmanageable, and owing to the peculiar nature of her cargo—had opened her " water-ways," where the deck and the sides were joined, about two inches. [When these part to any extent, a ship is always considered in a dangerous condition.] Then, again, one of the crew, a marine, had been literally frightened out of his wits, and remained crazy for some days.[1] One man had his arm fractured in two places, and another his leg broken.

---

[1] A fellow "marine" appears to have had more method in his madness. Whilst in the thick of the storm he hid himself in the bread-chest. He gave so much anxiety and trouble to find, that when found his friends were uncharitable enough to suggest that he had designs on the provisions.

THE ATLANTIC CABLE

The *Niagara*, on the other hand, had weathered the gale splendidly, though nevertheless with her (and still more with the smaller craft) it had been a hard and anxious time. She had lost her jib-boom, and the buoys she carried for suspending the cable had been washed from her sides, no man knew where.

After taking stock of things generally, a start was made to repair the damage.

The shifting of the upper part of the main coil on the *Agamemnon* into a hopeless tangle, entailed recoiling a considerable length of cable.

We will now once more continue our narrative in the words of Mr. Nicholas Woods, in reporting for *The Times* from the *Agamemnon*:—

"Neither Mr. Bright, nor Mr. Canning, nor Mr. Clifford was to be daunted by the aspect of a difficulty, however formidable. Absurd as the statement seemed at first, they were all positive that the tangle did not extend far down the coil, and they were right. Captain Preedy gave them his hearty assistance; men were at work day and night, drawing it out of the hold and coiling it aft on the main deck. For the first twenty-four hours the labour seemed hopeless, for so dense was the tangle that an hour's hard work would sometimes scarcely clear a half-mile. By-and-by, however, it began to mend, the efforts were redoubled, and late on Friday night 140 miles had been got out, and the remainder was found to be clear enough to commence work with. On the morning of Saturday, the 26th of June, all the preparations were completed for making the splice and commencing the great undertaking.

"The end of the *Niagara's* cable was sent on board the *Agamemnon*, the splice was made, a bent sixpence put in for luck, and at 2.50, Greenwich time, it was slowly lowered over the side, and disappeared for ever. The weather was cold and foggy, with a stiff breeze and dismal sort of sleet, and as there was no cheering or manifestation of enthusiasm of any kind, the whole ceremony had a most funereal effect, and seemed as solemn as if we were burying a marine, or some other mortuary task of the kind equally cheerful and enlivening. As it turned out, however, it was just as well that no display took place, as every one would have looked uncommonly silly when the same operation came to be repeated, as it had to be, an hour or so afterwards. It is needless making a long story longer, so I may state at once, that when each ship had payed out three miles or so, and they were getting well apart, the cable, which had been allowed to run too slack, broke on board the *Niagara*, owing to its overriding and getting off the pulley leading on to the machine.

"The break was, of course, known instantly :

both vessels put about and returned, a fresh splice was made, and again lowered over at half-past seven. According to arrangement 150 fathoms were veered out from each ship, and then all stood away on their course, at first at two miles an hour and afterwards at four. Everything then went well—the machine working beautifully, at 32 revolutions per minute—the screw at 26—the cable running out easily at five and five and a half miles an hour, the ship going four. The greatest strain upon the dynamometer was 2,500 lb., and this was only for a few minutes, the average giving only 2,000 lb. and 2,100 lb. At twelve at midnight, 21 nautical miles had been payed out, and the angle of the cable with the horizon had been reduced considerably. At about half-past three, 40 miles had gone, and nothing could be more perfect and regular than the working of everything, when suddenly, at 3.40 a.m., on Sunday, the 27th, Professor Thomson came on deck, and reported a total break of continuity; that the cable, in fact, had parted, and, as was believed at the time, from the *Niagara*. The *Agamemnon* was instantly stopped, and the breaks applied to the machinery, in order that the cable payed out might be severed from the mass in the hold, and so enable Professor Thomson to discover by electrical tests at about what distance from the ship the fracture had taken place.[1] Unfortunately, however, there

[1] By subsequent tests it was clear that at any rate the

was a strong breeze on at the time, with rather a heavy swell, which told severely upon the cable, and before any means could be taken to ease entirely the motion on the ship it parted, a few fathoms below the stern wheel, the dynamometer indicating a strain of nearly 4,000 lb. In another instant a gun and a blue light warned the *Valorous* of what had happened, and roused all on board the *Agamemnon* to a knowledge that the machinery was silent, and that the first part of the Atlantic cable had been laid, and effectually lost.

" The great length of cable on board both ships allowed a large margin for such mishaps as these, and the arrangement made before leaving England was that the splices might be renewed and the work recommenced till each ship had lost 250 miles of wire, after which they were to discontinue their efforts and return to Queenstown. Accordingly, after the breakage on Sunday morning the ships' heads were put about, and for the fourth time the *Agamemnon* again began the weary work of beating up against the wind for that everlasting rendezvous which we seemed destined to be always seeking. Apart from the regret with which all regarded the loss of the cable, there were other reasons for not wishing the cruise to be thus indefinitely prolonged, since there had

cable remaining on board was perfect. But after comparing notes with the *Niagara*, a strong belief was held that the cable probably parted at the bottom.

been a break in the continuity of the fresh
provisions, and for some days previously in the
ward-room the *pièces de resistance* had been inflam-
matory-looking *morceaux*, salted to an astonishing
pitch, and otherwise uneatable—for it was beef
which had been kept three years beyond its
warranty for soundness, and to which all were
then reduced.

"It was hard work beating up against the
wind; so hard, indeed, that it was not till the
noon of Monday, the 28th, that we again met the
Niagara; and, while all were waiting with im-
patience for her explanation of how she broke
the cable, she electrified every one by running up
the interrogatory, 'How did the cable part?'
This *was* astounding. As soon as the boats
could be lowered, Mr. Cyrus Field, with the electricians from the *Niagara*, came on board, and
a comparison of logs showed the painful and
mysterious fact, that at the same second of time
each vessel discovered that a total fracture had
taken place at a distance of certainly not less
than ten miles from each ship—as well as could
be judged, at the bottom of the ocean. The logs
on both sides were so clear as to the minute of
time, and as to the electrical tests showing not
merely leakage or defective insulations of the
wire, but a total fracture, that there was no room
left on which to rest a moment's doubt of the
certainty of this most disheartening fact. That

of all the many mishaps connected with the Atlantic telegraph this was the worst and most disheartening, since it proved that, after all that human skill and science can effect to lay the wire down with safety has been accomplished, there may be some fatal obstacles to success at the bottom of the ocean which can never be guarded against, for even the nature of the peril must always remain as secret and unknown as the depths in which it is to be encountered. Was the bottom covered with a soft coating of ooze in which it had been said the cable might rest undisturbed for years, as on a bed of down? or were there, after all, sharp-pointed rocks lying on that supposed plateau of Maury, Berryman and Dayman? These were the questions that some of those on board were asking.

" But there was no use in further conjecture, or in repining over what *had* already happened. Though the prospect of success appeared to be considerably impaired, it was generally considered that there was but one course left, and that was to splice again and make another—and what was fondly hoped would be a final—attempt. Accordingly, no time was lost in making the third splice, which was lowered over into 2,000 fathoms water at seven o'clock by ship's time the same night. Before steaming away, as the *Agamemnon* was now getting very short of coal, and the two vessels had some 100 miles of cable between them, it was agreed that if the wire parted again before the ships had gone each 100 miles from the rendezvous they were to return and make another splice; and as the *Agamemnon* was to sail back, the *Niagara*, it was decided, was to wait eight days for her reappearance. If, on the other hand, the 100 miles had been exceeded, the ships were not to return, but each make the best of its way to Queenstown. With this understanding the ships again parted, and, with the wire dropping steadily down between them, the *Niagara* and *Agamemnon* steamed away, and were soon lost in the cold, raw fog which had hung over the rendezvous ever since the operations had commenced.

" The cable, as before, payed out beautifully, and nothing could have been more regular and more easy than the working of every part of the apparatus. At first the ship's speed was only two knots, the cable going three, and three and a half with a strain of 1,500 lb., the horizontal angle averaging as low as seven, and the vertical about sixteen. By and by, however, the speed was increased to four knots, the cable going five, at a strain of 2,000 lb., and an angle of from twelve to fifteen. At this rate it was kept, with trifling variations, throughout almost the whole of Monday night, and neither Mr. Bright, Mr. Canning, nor Mr. Clifford ever quitted the machines for an instant. Towards the middle of the night, while the rate of the ship continued the same, the speed at which the cable payed-out slackened nearly a knot an hour, while the dynamometer indicated as low as 1,300 lb. This change could only be accounted for on the supposition that the water had shallowed to a considerable extent, and that the vessel was in fact passing over some submarine Ben Nevis or Skiddaw. After an interval of about an hour the strain and rate of progress of the cable again increased, while the increase of the vertical angle seemed to indicate that the wire was sinking down the side of a declivity. Beyond this there was no variation throughout Monday night, or, indeed, through Tuesday. The upper deck coil, which had weighed so heavily upon the ship—and still more heavily upon the minds of all during the past storms—was fast disappearing, and by twelve at midday on Tuesday, the 29th, 76 miles had been payed out to something like 60 miles progress of the ship. Warned by repeated failures, many of those on board scarcely dared to hope for success. Still the spirits of all rose as the distance widened between the ships. Things were going in splendid style—in such splendid style that "stock had gone up nearly 100 per cent." Those who had leisure for sleep were able to dream about cable-laying and the terrible effects of too great a strain. The first question which such as these ask on awakening is about the cable, and on being informed that it is all right, satisfaction ensues until the appearance of breakfast, when it is presumed this feeling is intensified. For those who do not derive any particular pleasure from the mere asking of questions, the harmonious music made by the paying-out machine during its revolutions supplies the information.

" Then, again, the electrical continuity—after all the most important item—was perfect, and the electricians reported that the signals passing between the ships were eminently satisfactory. The door of testing-room is almost always shut, and the electricians pursue their work undisturbed ; but it is impossible to exclude that spirit of scientific inquiry which will satiate its thirst for information even through a keyhole.

" Further, the weather was all that could be wished for. Indeed, had the poet who was so anxious for 'life on the ocean wave, and a home on the rolling deep' been aboard, he would have been absolutely happy, and perhaps even more desirous for a fixed habitation.

"The only cause that warranted anxiety was that it was evident the upper deck coil would be finished by about eleven o'clock at night, when the men would have to pass along in darkness the great loop which formed the communication between that and the coil in the main hold. This was most unfortunate, but the operation had been successfully performed in daylight during the experimental trip in the Bay of Biscay, and every precaution was now taken that no accident should occur. At nine o'clock by ship's time, when 146 miles had been payed out, and about 112 miles' distance from the rendezvous accomplished, the last flake but one of the upper deck coil came in turn to be used. In order to make it easier in passing to the main coil the revolutions of the screw were reduced gradually, by two revolutions at a time, from thirty to twenty, while the paying-out machine went slowly from thirty-six to twenty-two. At this rate, the vessel going three knots and the cable three-and-a-half, the operation was continued with perfect regularity, the dynamo-meter indicating a strain of 2,100 lb. Suddenly, without an instant's warning, or the occurrence of any single incident that could account for it, the cable parted, when subjected to a strain of less than a ton.[1] The gun that again told the *Valorous* of this fatal mishap brought all on board the *Agamemnon* rushing to the deck, for none could believe the rumour, that had spread like wildfire about the ship. But there stood the machinery, silent and motionless, while the fractured end of the wire hung over the stern wheel, swinging loosely to and fro. It seemed almost impossible to realise the fact that an accident so

[1] This was from the last turn in the coil, and subsequently it was discovered that, owing to the disturbance in the flooring of the tank during the storm, the cable had been damaged here.

instantaneous and irremediable should have occurred, and at a time when all seemed to be
going so well. Of course a variety of ingenious
suggestions were soon afloat, showing most satisfactorily how the cable must and ought to have
broken. There was a regular gloom that night
on board the *Agamemnon*, for from first to last
the success of the expedition had been uppermost
in the thoughts of all, and all had laboured for it
early and late, contending with every danger and
overcoming every obstacle and disaster that had
marked each day with an earnestness and devotion of purpose that is really beyond all praise.

"Immediately after the mishap, a brief consul-
tation was held by those in charge on board the
Agamemnon, and as it was shown that they had
only exceeded the distance from the rendezvous
by fourteen miles, and that there was still more
cable on board the two vessels than the amount
with which the original expedition last year was
commenced, it was determined to try for another
chance and return to the rendezvous, sailing there,
of course ; for Mr. Brown, the chief engineer, as
ultra zealous in the cause as a Board of Directors,
guarded the coal bunkers like a very dragon, lest,
if in coming to paying out the cable again, steam
should run short, thereby endangering the success
of the whole undertaking.

"For the fifth time, therefore, the *Agamemnon's*
head went about, and after twenty days at sea

she again began beating up against the wind for
the rendezvous, to try, if possible, to recommence
her labours. The following day the wind was
blowing strongly from the south-west, with mist
and rain, and Thursday, the 1st of July, gave
every one the most unfavourable opinion of July
weather in the Atlantic. The wind and sea were
both high—the wet fog so dense that one could
scarcely see the masts' heads, while the damp cold
was really biting. Altogether it was an atmosphere of which a Londoner would have been
ashamed even in November. Later in the day
a heavy sea got on; the wind increased without
dissipating the fog, and it was double-reefed topsails, and pitching and rolling as before. However, the upper deck coil of 250 tons being gone,
the *Agamemnon* was as buoyant as a lifeboat, and
no one cared how much she took to kicking
about, though the cold wet fog was a miserable
nuisance, penetrating everywhere, and making the
ship as wet inside as out. What made the
matter worse was that in such weather there
seemed no chance of meeting the *Niagara*—unless she ran into us, when cable-laying would
have gone on wholesale! In order to avoid such
a *contretemps*, and also to inform the *Valorous* of
our whereabouts, guns were fired, fog bells rung,
and the bugler stationed forward, to warn the
other vessels of our vicinity. Friday was the
ditto of Thursday, and Saturday worse than both

together, for it almost blew a gale, and there was a very heavy sea on. On Sunday, the 4th, it cleared, and the *Agamemnon*, for the first time during the whole cruise, reached the actual rendezvous, and fell in with the *Valorous*, which had been there since Friday, the 2nd, but the fog must have been even thicker there than elsewhere, for she had scarcely seen herself, much less anything else, till Sunday.

"During the remainder of that day and Monday, when the weather was very clear, both ships cruised over the place of meeting, but neither the *Niagara* nor *Gorgon* was there, though day and night the look out for them was constant and incessant. It was evident then that the *Niagara* had rigidly, but most unfortunately, adhered to the mere letter of the agreement regarding the 100 miles, and after the last fracture had at once turned back for Queenstown. On Tuesday, the 6th, therefore, as the dense fogs and winds set in again, it was agreed between the *Valorous* and *Agamemnon* to return once more to the rendezvous. But, as usual, the fog was so thick that the whole American navy might have been cruising there unobserved, so the search was given up, and at eight o'clock that night the ship's head was turned for Cork, and, under all sail, the *Agamemnon* at last stood homewards. The voyage home was made with ease and swiftness, considering the lightness of the wind, the trim of the ship, and that she only steamed three days, and at midday on Tuesday, July 12th, the *Agamemnon* cast anchor in Queenstown harbour, having met with more dangerous weather and encountered more mishaps than often falls to the lot of any ship in a cruise of thirty-three days."

Thus ended the most arduous and dangerous expedition that has ever been experienced in connection with cable-work. It, at any rate, had the advantage of supplying the public with some exciting reading in the columns of *The Times*, and Mr. Woods' graphic descriptions were much appreciated, even by other eye-witnesses!

As regards Charles Bright's diary during this period—with the constant strain of responsibility on his shoulders—it had necessarily consisted, with a few exceptions, of rough pencil notes referring to details such as miles run, cable paid out, strain on dynamometer, percentage of slack, etc.

He used to say that, arduous as it was, the life on board resembled a good ball—"the excitement keeping one going."

For purposes of accuracy it is to be regretted, of course, that those holding responsible positions have never time to write a record of the events, or even to attend to representatives of the press. If it were otherwise, there would be fewer false statements, which are passed on to posterity very often for want of being contradicted by the few who knew better but have other interests to attend to. In this particular instance, however, not only did Mr. Woods tell his story of the Atlantic cable-laying in a most palatable form—far more so than would be possible by any of the officials engaged in the work—but his account was notably accurate.

The *Niagara* had reached Queenstown as far back as July 5th. Those in charge, having found that they had run out a hundred and nine miles when "continuity" ceased, they considered that, in order to carry out their instructions, they should return at once to the above port, which they did. Before bearing homewards, however, and whilst the line was still hanging on to the ship's stern, opportunity was taken to make what proved to be an eminently satisfactory test in regard to the strength of the cable. After all hope of the continuity being restored was abandoned, the brakes were shut down so that the paying-out machine could not move. In this way the process of paying out was stopped for about an hour and a half, during which the whole weight of the *Niagara* was literally held by the slender cord, the wind blowing fresh all the time. And yet the cable did not break until the pressure put upon the brakes had reached an equivalent of over four tons strain!

On the two ships meeting at Queenstown discussion immediately took place (1) of course as to the cause of the cessation of continuity, and (2) regarding the course taken by the *Niagara* in returning home so promptly. The non-arrival of the *Agamemnon* till nearly a week later had been the cause of much alarm as regards her safety.

Section X

Finis coronat opus.

THE ATLANTIC TELEGRAPH CABLE WAS SUCCESSFULLY LAID, 5TH AUGUST, 1858.

Two mighty lands have shaken hands
  Across the deep wide sea ;
The world looks forward with new hope
  Of better times to be ;

276

THE ATLANTIC CABLE

For, from our rocky headland,
  Unto the distant West,
Have sped the messages of love
  From kind Old England's breast.

And from America to us
  Hath come the glad reply,
"We greet you from our heart of hearts,
  We hail the new-made tie ;
We pledge again our loving troth
  Which under Heaven shall be
As stedfast as Monadnoc's cliffs,
  And deep as is the sea!"

Henceforth the East and West are bound
  By a new link of love,
And as to Noah's ark there came
  The olive-bearing dove,
So doth this ocean telegraph,
  This marvel of our day,
Give hopeful promise that the tide
  Of war shall ebb away.

No more, as in the days of yore,
  Shall mountains keep apart,
No longer oceans sunder wide
  The human heart from heart,
For man hath grasped the thunderbolt,
  And made of it a slave
To do its errands o'er the land,
  And underneath the wave.

Stretch on, thou wonder-working wire !
  Stretch North, South, East and West,
Deep down beneath the surging sea,
  High o'er the mountain's crest.

SIR CHARLES TILSTON BRIGHT

Stretch onwards without stop or stay,
  All lands and oceans span,
Knitting with firmer, closer bonds
  Man to his brother man.

Stretch on, still on, thou wondrous wire!
  Defying space and time,
Of all the mighty works of man
  Thou art the most sublime.
On thee, bright-eyed and joyous Peace,
  Her sweetest smile hath smiled,
For, side by side, thou bring'st again
  The mother and the child.

Stretch on! O may a blessing rest
  Upon this wondrous deed,
This conquest where no tears are shed,
  In which no victims bleed!
May no rude storm disturb thy rest
  Nor quench the swift-winged fire
That comes and goes at our command
  Along thy wondrous wire.

Long may'st thou bear the messages
  Of love from shore to shore,
And aid all good men in the cause
  Of Him whom we adore:
For thou art truly but a gift
  By the All-bounteous given;
The minds that thought, the hands that wrought,
  Were all bestowed by Heaven.
          —*The British Workman.*

The sad tale of disaster commenced to spread

abroad immediately on the *Niagara's* arrival in Queenstown; and when Mr. Field hastened to London to meet the other Directors of the Company, he found that the news had not only preceded him but had already had its effect. The Board was soon called together. It met in the same room in which, six weeks earlier, it had discussed the prospects of the expedition with full confidence of success. Now it met as a council of war summoned after a terrible defeat, to decide whether to surrender or to try once more the chances of battle.

Says Field:—"Most of the directors looked blankly in one another's faces." With some the feeling was one akin to despair. It was thought by many that there was nothing left on which to found an expectation of future success, or to encourage the expenditure of further capital upon an adventure so "completely visionary."

Sir William Brown (the first chairman) whilst recommending complete abandonment of the undertaking, suggested "a sale of the cable remaining on board the ships, and a distribution of the proceeds amongst the shareholders."

Mr. Brooking, the vice-chairman, also now convinced of the impracticability of the undertaking, sent in his resignation.

Bolder counsels, however, were destined to prevail. There were those who thought there was still a chance, like Robert Bruce who, after twelve battles and twelve defeats, yet believed that a thirteenth *might* bring victory.

Besides the projectors—J. W. Brett, Charles Bright, Cyrus Field, and Wildman Whitehouse— Mr. Curtis Lampson (who succeeded Mr. Brooking as deputy-chairman) made a firm stand for action at once, as did also Professor Thomson.[1] These advocates of non-surrender at length succeeded in carrying an order for the immediate sailing of the Expedition for a final effort. It was this effort which proved to the world the possibility of telegraphing from one hemisphere to the other.

The order to advance having been given, the ships immediately took in coal and other necessaries.

During this interval, and whilst in London, Charles Bright took an opportunity of running down to his Harrow home for a single day and night, to catch a glimpse of those dearest to him.

On leaving, he remarked to his young wife, " I don't say we shall do it even this time, but we shall do it *some* time." This remark was very characteristic of the man. It will probably be admitted that the failing with many is that though they set their teeth at a thing they do not do so

---

[1] Whilst the ships were lying at Queenstown, Professor Thomson had transmitted signals through the entire length of cable on the two ships, thereby again demonstrating the electrical practicability of the line.

for long enough. This could scarcely be said of the subject of our biography.

When everything and everybody had been shipped, the squadron left Queenstown once more on Saturday, July 17th. As the ships sailed out of the harbour of Cork, it was with none of the enthusiasm which attended their departure from Valentia the year before, or even the small amount excited when leaving Plymouth on June 10th. Nobody so much as cheered! In fact, their mission was by this time spoken of as a "mad freak" of "stubborn ignorance!"

The squadron was the same as on the last occasion. It was agreed that the ships should not attempt to keep together this time, but that each should make its way to the given latitude and longitude.

The staff were composed and berthed as before, Mr. Field once more taking up his quarters aboard the *Niagara*. Moreover, the expedition was again accompanied by the same literary talent, and we cannot do better now than give the story as it is continued by Mr. Nicholas Woods on behalf of The *Times*,[1] so far as the *Agamemnon* (containing Charles Bright) is concerned.

As your readers have already been informed by

---

[1] *The Times*, Wednesday August 11th, 1858.

telegraph, the submarine communication between the Old and New Worlds is now an accomplished fact. In the face of difficulties and dangers—the magnitude of which cannot be properly appreciated by those not engaged in the work—the engineers engaged in this undertaking, have, with almost untiring energy, adhered to their all but hopeless task with that perseverance which is sure, sooner or later, to lead to success. There were but few some twenty days ago who, after the unsuccessful return of the squadron to Queenstown, would have dared to predict such a speedy and glorious termination to all the trials and difficulties that the promoters of this undertaking have undergone. The final accomplishment of the scheme seemed indeed up to the last moment to hang upon a hair. Many serious difficulties had to be encountered during the six days and a half that the operations lasted ; any one of which, had not chance favoured us, might have ruined the expedition and delayed the advance of ocean telegraphs perhaps more than half a century. But the difficult task has now been accomplished, and it only remains for us to accept the benefits which it will undoubtedly confer upon the community. Wonderful as the conception of conveying sensations from continent to continent, across the almost unknown depths of the ocean, may seem to us now, yet in a very little time people will forget the marvel while profiting by the fact ; and, without remembering the years of anxious toil and discouragement which those who have secured this boon to the community have undergone to secure success, the wonder will be, not that the undertaking has been carried out at all, but that it had not been accomplished long before. It has been the custom of mankind to honour the lives and celebrate the deeds of great statesmen, successful warriors, and eminent divines. Indeed, of such materials are the links in the chain of history chiefly composed. But those men who, by patient thought and persevering action, have achieved victories over matter, which secure to the community permanent advantages, very often have their trouble for their reward. It is to be hoped that this may not be the case with those who have been mainly instrumental in bringing this great scheme to a successful termination. It must be confessed that the prospects of success were very remote when the squadron left Queenstown on the 17th of last month. The amount of cable in the two ships had been reduced by nearly 400 miles ; and the recollection of three separate and most unaccountable breakages was still fresh in the minds of all who had accompanied the first expedition. There was no reason whatever for supposing that the very same thing might not occur again. The cable might, and evidently did, as far as the contractors are concerned, fulfil all the guaranteed requirements ; and the numerous accidents which occurred might be due to the cable having become injured during the gale. This supposition, though it may be gratifying to Messrs. Glass, Elliot & Co., and to Messrs. Newall & Co., was no consolation to either the engineers or the shareholders. Under these circumstances, it is not surprising that many regarded the prosecution of the scheme as a waste of the shareholders' money. However—in spite of the most vehement opposition—the majority of the directors determined to despatch the expedition to try their fortune again in mid ocean before they abandoned the scheme altogether as impracticable. Accordingly, on the morning of Saturday, the 17th of July, the *Valorous, Gorgon,* and *Niagara,* having completed coaling, steamed away from Queenstown for the *rendezvous.* The *Agamemnon,* having to wait for Professor W. Thomson, one of the directors, who took charge of the electrical department on board,[1] did not weigh anchor until 2 o'clock on the following morning.

As the ships left the harbour there was apparently no notice taken of their departure by those on shore or in the vessels anchored around them. Every one seemed impressed with the conviction that we were engaged in a hopeless enterprise; and the squadron seemed rather to have slunk away on some discreditable mission, than to have sailed for the accomplishment of a grand national scheme. It was just dawn when the *Agamemnon* got clear of Queenstown harbour; but, as the wind blew stiff from the south-west, it was nearly 10 o'clock before she rounded the Old Head of Kinsale, a distance of only a few miles. The weather remained fine during the day; and, as the *Agamemnon* skirted along the wild and rocky shore of the south-west coast of Ireland, those on board had an excellent opportunity of seeing the stupendous rocks which rise from the water in the most grotesque and fantastic shapes. About 5 o'clock in the afternoon Cape Clear was passed; and though the coast gradually edged away to the northward of our course, yet it was nearly dark before we lost sight of the rocky mountains which surround Bantry Bay and the shores of the Kenmare river. By Monday, the 19th, we had left the land far behind us, and thence fell into the usual dull monotony of sea life.

Of the voyage out, there is little to be said. It was not checkered by the excitement of continual storms or the tedium of perpetual calms, but we had a sufficient admixture of both to render our passage to the *rendezvous* a

---

[1] The gentleman holding the position of electrician to the Company, Mr. Whitehouse, was still, under medical advice, prevented from accompanying the expedition.

very ordinary and uninteresting one. For the first week the barometer remained unusually low, and the numbers of those natural barometers, Mother Carey's chickens, that kept in our wake kept us in continual expectation of heavy weather. With very little breeze, or wind, the screw was got up and sail made, so as to husband our coals as much as possible; but it generally soon fell calm and obliged Captain Preedy reluctantly to get up steam again. In consequence of continued delays and changes from steam to sail, and from sail to steam again, much fuel was expended, and not more than eighty miles of distance made good each day. On Sunday, the 25th, however, the weather changed, and for several days in succession there was an uninterrupted calm. The moon was just at the full; and for several nights it shone with a brilliancy which turned the smooth sea into one silvery sheet, which brought out the dark hull and white sails of the ship in strong contrast to the sea and sky as the vessel lay all but motionless on the water—the very impersonation of solitude and repose. Indeed, until the *rendezvous* was gained, we had such a succession of beautiful sunrises, gorgeous sunsets, and tranquil moonlight nights, as would have excited the most enthusiastic admiration of any one but persons situated as we were. But by us, such scenes were regarded only as the annoying indications of the calm which delayed our progress and wasted our coals. To say that it was calm is not doing full justice to it— there was not a breath in the air, and the water was as smooth as a mill-pond. Even the wake of the ship scarce ruffled its surface; and the gulls, which had visited us almost daily, and to which our benevolent liberality had dispensed innumerable pieces of pork, threw an almost unbroken shadow upon it as they stooped in their flight to pick up the largest and most tempting. It was generally remarked that cable-laying under such circumstances would be mere child's play.

In spite of the unusual calmness of the weather in general, there were days on which our former unpleasant experiences of the Atlantic were brought forcibly to our recollection—when it blew hard, and the sea ran sufficiently high to reproduce on a minor scale some of the discomforts of which the previous cruise had been so fruitful. These days, however, were the exception and not the rule ; and served to show how much more pleasant was the inconvenient calm than the weather which had previously prevailed.

The precise point of the *rendezvous*—marked by a dot on the chart—was reached on the evening of Wednesday, the 28th July, just eleven days after our departure from Queenstown. The voyage out was a lazy one. Now things are different, and we no longer hear of the prospects of the heroes and heroines of the romances and novels which have formed the staple food for animated discussion for some days past. The rest of the squadron were in sight at nightfall, but at such a considerable distance that it was past 10 o'clock on the morning of Thursday, the 29th, before the *Agamemnon* joined them. Some time previous to reaching the *rendezvous* the engineer-in-chief (Mr. Bright) went up in the shrouds on the look-out for the other ships, and accordingly had to "pay his footing"—much to the amusement of his staff. Most of them being more advanced in years would not probably have been so equal to the task in an athletic sense.

After the ordinary laconic conversation which characterise code flag signals,[1] we were as usual greeted by

---

[1] Such as, "I hope you are all well." "Very well,

a perfect storm of questions as to what kept us so much behind our time, and learned that all had come to the conclusion that the ship must have got on shore on leaving Queenstown harbour. The *Niagara*, it appeared, had arrived at the *rendezvous* on Friday night, the 23rd, the *Valorous* on Sunday, the 25th, and the *Gorgon* on the afternoon of Tuesday, the 27th.

The day was beautifully calm; so no time was to be lost before making the splice in lat. 52° 9′ N., long. 32° 27′ W., and soundings of 1,500 fathoms. Boats were soon lowered from the attendant ships; the two vessels made fast by a hawser, and the *Niagara's* end of the cable conveyed on board the *Agamemnon*. About half-past 12 o'clock the splice was effectually made; but with a very different frame from carefully rounded semi-circular boards which had been used to enclose the junctions on previous occasions.[1] It consisted merely of two straight boards hauled over the joint and splice,[2] with the iron rod and

I thank you." A touch of irony characterised one, however, when the *Gorgon* asked the *Niagara* if she had any coal to spare, the reply—this time by word of mouth—came, "None at all. I think the *Agamemnon* could give you some as she can't have burned much since she left."

[1] The actual form of this apparatus was shown on p. 156, amongst the arrangements made by Charles Bright for the first (1857) expedition.

[2] This was of the order sometimes described as a ball-splice.

leaden plummet attached to the centre. In hoisting it out from the side of the ship, however, the leaden sinker broke short off and fell overboard. There being no more convenient weight at hand a 32-lb. shot was fastened to the splice instead; and the whole apparatus was quickly dropped into the sea without any formality—and, indeed, almost without a spectator—for those on board the ship had witnessed so many beginnings to the telegraphic line that it was evident they despaired of there ever being an end to it.

The stipulated 210 fathoms of cable having been paid out to allow the splice to sink well below the surface, the signal to start was hoisted, the hawser cast loose, and the *Niagara* and *Agamemnon* start for the last time at about 1 p.m. for their opposite destinations.

The announcement comes from the electrician's testing-room that the continuity is perfect, and with this assurance the engineers go on more boldly with the work. In point of fact the engineers may be said to be very much under the control of the electricians during paying out; for if they report anything wrong with the cable the engineers are brought to a stand until they are allowed to go on with their operations by the announcement of the electricians that the insulation is perfect and the continuity all right. The testing room is where the subtle current which flows along the conductor is generated; and where the mysterious apparatus by which electricity is weighed and measured—as a marketable commodity—is fitted up. The system of testing and of transmitting and receiving signals through the cable from ship to ship during the process of paying out must now be briefly referred to. It consists of an exchange of currents sent alternately every ten minutes by each ship. These not only serve to give an accurate test of the continuity and insulation of the conducting wire from end to end, but also to give certain signals which it is desirable to send for information purposes. For instance, every ten miles of cable paid out is signalized from ship to ship, as also the approach to land or momentary stoppage for splicing, shifting to a fresh coil, etc. The current in its passage is made to pass through an electromagnetometer,[1] an instrument used by Mr. Whitehouse.[2] It is also conveyed in its passage at each end of the cable through the reflecting galvano-

[1] Though bearing this somewhat cumbersome and elaborate title this instrument was practically nothing more nor less than an ordinary "detector," its capacity for actually *measuring* the current being of an extremely limited character.

[2] This was fully described in Mr. Whitehouse's Cheltenham B. A. paper (see the *Athenæum* of September, 1856). That gentleman's complete apparatus was also described in the *Engineer* of January, 1857, a journal which had made its first appearance exactly a year previously.

meter and speaking instrument just invented by Prof. Thomson; and it is this latter which is so invaluable, not only for the interchange of signals, but also for testing purposes. The deflections read on the galvanometer, as also the degree of charge and discharge indicated by the magnetometer, are carefully recorded. Thus, if a defect of continuity or insulation occurs it is brought to light by comparison with those received before.

For the first three hours the ships proceeded very slowly, paying out a great quantity of slack; but after the expiration of this time the speed of the *Agamemnon* was increased to about five knots, the cable going at about six, without indicating more than a few hundred pounds of strain upon the dynamometer.

Shortly after 4 o'clock a very large whale was seen approaching the starboard bow at a great speed, rolling and tossing the sea into foam all round; and for the first time we felt a possibility for the supposition that our second mysterious breakage of the cable might have been caused after all by one of these animals getting foul of it under water. It appeared as if it were making direct for the cable; and great was the relief of all when the ponderous living mass was seen slowly to pass astern, just grazing the cable where it entered the water—but fortunately without doing any mischief.

All seemed to go well up to about eight o'clock;

the cable paid out from the hold with an even-
ness and regularity which showed how carefully
and perfectly it had been coiled away. The
paying-out machine also worked so smoothly that
it left nothing to be desired. The brakes are
properly called self-releasing ; and although they

IN COLLISION WITH A WHALE DURING CABLE LAYING.

can by means of additional weights be made to
increase the pressure or strain upon the cable,
yet, until these weights are still further increased
(at the engineer's instructions) it is impossible to
augment the strain in any other way. To guard
against accidents which might arise in consequence
of the cable having suffered injury during the
storm, the indicated strain upon the dynamometer

was never allowed to go beyond 1,700 lb., or less than one-quarter what the cable is estimated to bear. Thus far everything looked promising.

But in such a hazardous work no one knows what a few minutes may bring forth, for soon after eight o'clock an injured portion of the cable[1] was discovered about a mile or two from the portion paying out. Not a moment was lost by Mr. Canning, the engineer on duty, in setting men to work to cobble up the injury as well as time would permit, for the cable was going out at such a rate that the damaged portion would be paid overboard in less than twenty minutes, and former experience had shown us that to check either the speed of the ship or the cable would, in all probability, be attended by the most fatal results. Just before the lapping was finished Professor Thomson reported that the electrical continuity of the wire had ceased, but that the insulation was still perfect. Attention was naturally directed to the injured piece as the probable source of the stoppage, and not a moment was lost in cutting the cable at that point, with the intention of making a perfect splice.[2] To the

[1] This was some of the cable damaged during the storm, like that which had broken at the end of the previous attempt. The bottom of the hold here was found afterwards to be in a very disordered state.

[2] In connection with the above, an extract from Bright's diary will serve to fill up some gaps :—

consternation of all, the electrical tests applied showed the fault to be overboard, and in all probability some fifty miles from the ship.

Not a second was to be lost, for it was evident that the cut portion must be paid overboard in a few minutes; and in the meantime the tedious and difficult operation of making a splice had to be performed. The ship was

"29th July, Greenwich time, 10 p.m. Signals ceased from *Niagara*. Professor Thomson reported loss of continuity, with insulation good. To ascertain whether fault was at the piece of cable which was about to be lapped, the cable was sprung open at this point and the gutta percha wire pricked, and the part in the ship found good ; pricked again nearer stern, found good inside ship.

"No indication of fracture during the time. It was then cut about ten turns from the outgoing part, and the test showed the loss of continuity to be far from the ship —probably more than 40 miles, but decidedly less than 200.

"Joint made again as quickly as possible, and tested. Want of continuity and good insulation still experienced. When one turn off joint, commenced veering out again. Ship's time, 9.5 p.m. Splice paid over safely. Same results. Strong current came. On testing, "earth" found about middle of cable, and on currents again coming it was concluded that the cable had been cut on board the *Niagara*.

"Signals then sent and received regularly, and showed 1,200 or 1,300 miles in circuit."

Note.—This trouble might have been avoided had complete speaking arrangements been made.

immediately stopped, and no more cable paid out than was absolutely necessary to prevent it breaking. As the stern of the ship was lifted by the waves a scene of the most intense excitement followed. It seemed impossible, even by using the greatest possible speed, and paying out the least possible amount of cable, that the junction could be finished before the part was taken out of the hands of the workmen. The main hold presented an extraordinary scene. Nearly all the officers of the ship, and of those connected with the expedition, stood in groups about the coil, watching with intense anxiety the cable, as it slowly unwound itself nearer and nearer the joint, while the workmen worked at the splice as only men could work who felt that the life and death of the expedition depended upon their rapidity. But all their speed was to no purpose, as the cable was unwinding within a hundred fathoms; and, as a last and desperate resource, the cable was stopped altogether, and for a few minutes the ship hung on by the end. Fortunately, however, it was only for a few minutes, as the strain was continually rising above two tons, and it would not hold on much longer. When the splice was finished the signal was made to loose the stoppers, and it passed overboard in safety.

When the excitement, consequent upon having so narrowly saved the cable, had passed away, we awoke to the consciousness that the case was yet as hopeless as ever, for the electrical continuity was still entirely wanting.

Preparations were consequently made to pay out as little rope as possible, and to hold on for six hours in the hope that the fault, whatever it might be, might mend itself, before cutting the cable and returning to the rendezvous to make another splice. The magnetic needles on the receiving instruments were watched closely for the returning signals; when, in a few minutes, the last hope was extinguished by their suddenly indicating dead earth, which tended to show that the cable had broken from the *Niagara*, or that the insulation had been completely destroyed.

Nothing, however, could be done. The only course was to wait until the current should return or take its final departure. And it *did* return—with greater strength than ever—for in three minutes every one was agreeably surprised by the intelligence that the stoppage had disappeared, and that the signals had again appeared at their regular intervals from the *Niagara*.[1] It is needless to say what a load of anxiety this news removed from the minds of every one; but

---

[1] Later on it was made clear that this mysterious temporary want of continuity, accompanied by an apparent variation in the insulation was due to a defect in the more or less inconstant sand battery used aboard the latter vessel.

the general confidence in the ultimate success of the operations was much shaken by the occurrence, for all felt that every minute a similar accident might occur.[1]

---

[1] This unpleasant incident regarding the continuity was never forgotten to the last, and forbade all to indulge in sanguine expectations even when prospects seemed clear. Mr. Mullaly says, in his diary: " The sailors, who are somewhat in the dark as to the scientific definition of the term 'continuity,' believe it to be at the bottom of all the trouble, and credit it even with vindictive qualities. 'Darn the continuity,' said an old 'salt,' after what was to him a highly scientific but to his audience of messmates a rather foggy dissertation on the subject of cable work. 'Darn the continuity; I wish they would get rid of it altogether. It has caused a darned sight more trouble than the business is worth. I say they ought to do without it, and let it go. I believe they'd get the cable down if they didn't pay any attention to it. You see,' he went on, ' I was on the last exhibition (*expedition* he meant, but it was all the same—his messmates did not mistake his meaning) and I thought I'd never hear the end of it. They were always talking about it, and one night when we were out last year it was gone for two hours, and we thought that was the end of the affair, and we should never hear of it again. But it came back, and soon after the cable busted. Now, I tell you what, men, I'll never forget the night, I tell you. We all felt we had lost our best friend. After that I have never heard the word continuity, or contiguity, or whatever it is, mentioned, but I was always afraid something was going to happen. And that's a fact.' This was conclusive on the

THE ATLANTIC CABLE

For some time the paying out continued as usual, but towards the morning another damaged place was discovered in the cable. There was fortunately time, however, to repair it in the hold without in any way interfering with the operations, beyond for a time slightly reducing the speed of the ship. During the morning of Friday, the 30th, everything went well. The ship had been kept at the speed of about five knots, the cable going out at six, the average angle with the horizon at which it left the ship being about 15°, while the indicated strain upon the dynamometer seldom showed more than 1,600 lb. to 1,700 lb.

Observations made at noon showed that we had made good 90 miles from the starting point since the previous day, with an expenditure —including the loss in lowering the splice and during the subsequent stoppages—of 135 miles of cable. During the latter portion of the day the barometer fell considerably, and towards the evening it blew almost a gale of wind from the eastward, dead ahead of our course. As the breeze freshened the speed of the engines was

minds of the majority of his hearers. However, a number were of opinion that it was all right, and—at the risk of being considered humbugs—asserted their belief that whatever might be said against the continuity they couldn't do without it, and that, on the contrary, it was because it was gone all the trouble had occurred."

gradually increased, but the wind more than increased in proportion, so that before the sun went down the *Agamemnon* was going full steam against the wind, only making a speed of about four knots.

During the evening topmasts were lowered, and spars, yards, sails, and indeed, everything aloft that could offer resistance to the wind was sent down on deck. Still' the ship made but little way, chiefly in consequence of the heavy sea; though the enormous quantity of fuel consumed showed us that if the wind lasted we should be reduced to burning the masts, spars, and even the decks, to bring the ship into Valentia. It seemed to be our particular ill-fortune to meet with head winds whichever way the ship's head was turned. On our journey out we had been delayed and obliged to consume an undue proportion of coal for want of an easterly wind, and now all our fuel was wanted because of one. However, during the next day the wind gradually went round to the south-west, which, though it raised a very heavy sea, allowed us to husband our small remaining store of fuel.

At noon on Saturday, the 31st of July, observations showed us to be in lat. 52° 23′ N., and long. 26° 44′ W., having made good 120 miles of distance since noon of the previous day, with a loss of about 27 per cent. of cable.

THE ATLANTIC CABLE

The *Niagara*, as far as could be judged from the amount of cable she paid out—which by a previous arrangement was signalled at every ten miles—kept pace with us within one or two miles the whole distance across.

During the afternoon of Saturday the wind again freshened up, and before nightfall it blew nearly a gale of wind, and a tremendous sea ran before it from the south-west, which made the *Agamemnon* pitch and toss to such an extent that it was thought impossible the cable could hold through the night. Indeed, had it not been for the constant care and watchfulness exercised by Mr. Bright and the two energetic engineers, Mr. Canning and Mr. Clifford, who acted with him, it could not have been done at all. Men were kept at the wheels of the machine to prevent their stopping (as the stern of the ship rose and fell with the sea), for had they done so, the cable must undoubtedly have parted.[1] During

---

[1] The paying-out apparatus was roped-in, with a notice placed conspicuously, reading thus : "No one here except the Engineer's Watch." This was certainly laconic ; but if any other than the privileged few made his way inside the sacred ground, the marine who stood close by informed him he must leave. This was not all, however, for if under the impression that he was at liberty to talk to the operator in charge of the dynamometer, he was soon made aware of the absurdity of such an idea by another inscription to the effect that no conversation was allowed with that particular individual.

Sunday the sea and wind increased, and before the evening it blew a smart gale.

Now, indeed, were the energy and activity of all engaged in the operation tasked to the utmost. Mr. Hoar and Mr. Moore, the two engineers who had the charge of the relieving wheels of the dynamometer, had to keep watch and watch alternately every four hours ; and while on duty, durst not let their attention be removed from their occupation for one moment, for on their releasing the breaks every time the stern of the ship fell into the trough of the sea entirely depended the safety of the cable, and the result shows how ably they discharged their duty.

Throughout the night there were few who had the least expectation of the cable holding on till morning, and many lay awake listening for the sound that all most dreaded to hear—viz., the gun which should announce the failure of all our hopes. But still the cable—which, in comparison with the ship from which it was paid out and the gigantic waves among which it was delivered was but a mere thread—continued to hold on, only leaving a silvery phosphorus line upon the stupendous seas as they rolled on towards the ship.

With Sunday morning came no improvement in the weather; still the sky remained black and stormy to windward, and the constant violent squalls of wind and rain which prevailed during the whole day, served to keep up, if not to augment, the height of the waves.

But the cable had gone through so much during the night that our confidence in its continuing to hold was much restored.[1] At noon observations showed us to be in lat. 52° 26′ N , and long. 23° 16′ W., having made good 130 miles from noon of the previous day, and about 350 from our starting point in mid ocean. We had passed by the deepest sounding of 2,400 fathoms, and over more than half of the deep water generally, while the amount of cable still remaining in the ship was more than sufficient to carry us to the Irish coast, even supposing the continuance of the bad weather should oblige us to pay out nearly the same amount of slack cable as hitherto.

Thus far things looked very promising for our ultimate success. But former experience showed us only too plainly that we could never suppose that some accident might not arise until the ends had been fairly landed on the opposite shores.

One of the expedition made some notes in the present-tense-story form, which are reproduced here as indicative of the feelings indulged by those on board about this time :—

The cable is the absorbing subject of conversation. We hardly dare ask ourselves if we shall lay the line the

[1] A note in Bright's rough diary says :—" 8 a.m., insulation reported better than ever."

whole distance—it seems too much to hope for—and we dread to think of the future. We count each day, not by hours, but by minutes. The sound of the machinery has become as familiar to us as that of our own voices ; and when it is drowned in any other noise, we listen with eagerness to hear it again. The barometer is consulted hourly, and its variations watched with a jealous eye, for we can now appreciate how much depends on the weather. The sight of that thread-like wire battling with the wind and sea, produces a feeling somewhat akin to that with which you would watch the struggles of a drowning man, whom you have not the power of assisting. There is a strong undercurrent of confidence, though we are still some way from the end ; and a kink in the cable, or a hole running through the gutta percha into the conductor—a tiny hole, such as you could not force a hair through—would render the labour of years utterly unavailing.

That group of sailors near the cook's galley are engaged in an animated discussion on the all-prevailing topic. One of the number is trying to persuade his messmates that it is impossible to lay it ; but they lend him rather unwilling ears.

Altogether the cable is getting into better repute, and specimens of it are more highly prized than they were before. Nothing is thought of during the day but the cable, and I believe two-thirds of the crew don't *dream* of anything else. Some of us are unreasonable enough to wish that things were still better, and that we were once more at home and amongst our friends—in fact that this terrible struggle between hope and fear were at an end. Then our thoughts turn to the scene of wild excitement ashore, when it is learnt that the " impracticable enterprise" has, after all, succeeded—that is to say, if all continues to go on well to the finish.

THE ATLANTIC CABLE

During Sunday night and Monday morning the weather continued as boisterous as ever. It was only by the most indefatigable exertions of the engineer upon duty that the wheels could be prevented from stopping altogether as the vessel rose and fell with the sea; and once or twice they did come completely to a standstill, in spite of all that could be done to keep them moving. Fortunately, however, they were again set in motion before the stern of the ship was thrown up by the succeeding wave. No strain could be placed upon the cable of course; and though the dynamometer occasionally registered 1,700 lb. as the ship lifted, it was oftener below 1,000 lb., and was frequently nothing, the cable running out as fast as its own weight and the speed of the ship could draw it. But, even with all these forces acting unresistingly upon it, the cable never paid itself out at a greater speed than eight knots, at the time the ship was going at the rate of six knots and a half. Subsequently, however, when the speed of the ship even exceeded six knots and a half, the cable never ran out so quick. The average speed maintained by the ship up to this time—and, indeed for the whole voyage—was about five knots and a half, the cable, with occasional exceptions, running some 30 per cent. faster.

At noon on Monday, August the 2nd, observations showed us to be in lat. 52° 35′ N; long. 19° 48′ W. Thus we had made good $127\frac{1}{2}$

miles since noon of the previous day, and had completed more than half-way to our ultimate destination.

During the afternoon, an American three-masted schooner, which afterwards proved to be the *Chieftain*, was seen standing from the eastward towards us. No notice was taken of her at first, but when she was within about half a mile of the *Agamemnon*, she altered her course and bore right down across our bows. A collision, which might prove fatal to the cable, now seemed inevitable; or could only be avoided by the equally hazardous expedient of altering the *Agamemnon's* course. The *Valorous* steamed ahead, and fired a gun for her to heave to, which, as she did not appear to take much notice of, was quickly followed by another from the bows of the *Agamemnon*, and a second and third from the *Valorous*. But still the vessel held on her course; and, as the only resource left to avoid a collision, the course of the *Agamemnon* was altered just in time to pass within a few yards of her. It was evident that our proceedings were a source of the greatest possible astonishment to them, for all her crew crowded upon her deck and rigging. At length they evidently discovered who we were and what we were doing, for the crew manned the rigging, and dipping the ensign several times, they gave us three hearty cheers. Though the *Agamemnon* was obliged to acknowledge these congratulations in due form, the feelings of annoyance with which we regarded the vessel, which (either by the stupidity or carelessness of those on board) was so near adding a fatal and unexpected mishap to the long chapter of accidents which had already been encountered, may easily be imagined.

To those below—who of course did not see the ship approaching—the sound of the first gun came like a thunderbolt, for all took it as a signal of the breaking of the cable. The dinner tables were deserted in a moment, and a general rush made up the hatches to the deck; but before reaching it their fears were quickly banished by the report of the succeeding gun, which all knew well could only be caused by a ship in our way, or a man overboard.

Throughout the greater part of Monday morning the electrical signals from the *Niagara* had been getting gradually weaker, until they ceased altogether for nearly three-quarters of an hour. Then Professor Thomson sent a message to the effect that the signals were too weak to be read; and, in a little while, the deflections returned even stronger than they had ever been before. Towards the evening, however, they again declined in force for a few minutes.[1]

[1] In connection with the above, Bright's diary says :—
" Aug. 2, 1.40 p.m. Prof. Thomson reports no regular signals from the *Niagara* for three terms of the usual ten minutes. Currents come, but no intelligible signals accord-

With the exception of these little stoppages, the electrical condition of the submerged wire seemed to be much improved. It was evident that the low temperature of the water at the immense depth improved considerably the insulating properties of the gutta percha, while the enormous pressure to which it must have been subjected probably tended to consolidate its texture, and to fill up any air bubbles or slight faults in manufacture which may have existed.

The weather during Monday night moderated a little; but still there was a very heavy sea on, which endangered the wire every second minute.

About three o'clock on Tuesday morning all on board were startled from their beds by the loud booming of a gun. Every one—without waiting for the performance of the most particular toilet—rushed on deck to ascertain the cause of the disturbance. Contrary to all expectation, the cable was safe; but just in the grey light could be seen the *Valorous*—rounded to in the most warlike attitude, firing gun after gun in quick succcession towards a large American barque, which, quite unconscious of our proceedings, was standing

ing to the arranged methods. It is possible they may be earth currents."

It subsequently transpired that the trouble had been due to a fault in the *Niagara's* ward-room coil. As soon as the electricians discovered this, and had it cut out, all went smoothly again.

right across our stern. Such loud and repeated remonstrances from a large steam frigate were not to be despised ; and evidently, without knowing the why or the wherefore, she quickly threw her sails aback, and remained hove to. Whether those on board her considered that we were engaged in some filibustering expedition, or regarded our proceedings as another British outrage upon the American flag, it is impossible to say ; but certain it is that—apparently in great trepidation—she remained hove to until we had lost sight of her in the distance.

Tuesday was a much finer day than any we had experienced for nearly a week, but still there was a considerable sea running, and our dangers were far from passed ; yet the hopes of our ultimate success ran high. We had accomplished nearly the whole of the deep portions of the route in safety, and that, too, under the most unfavourable circumstances possible ; therefore there was every reason to believe that, unless some unforeseen accident should occur, we should accomplish the remainder. Observations at noon placed us in lat. 5° 26′ N. ; long. 16° 7′ 40″ W.; having run 134 miles since the previous day.

About five o'clock in the evening the steep submarine mountain which divides the steep telegraphic plateau from the Irish coast was reached ; and the sudden shallowing of the water had a very marked effect upon the cable, causing the strain, and the speed, to lessen every minute. A great deal of slack was paid out,[1] to allow for any greater inequalities which might exist, though undiscovered by the sounding line.

About ten o'clock the shoal water of 250 fathoms was reached. The only remaining anxiety now was the changing from the lower main coil to that upon the upper deck; and this most dangerous operation was successfully performed between three and four o'clock on Wednesday morning.

Wednesday was a beautifully calm day; indeed it was the first on which any one would have thought of making a splice since the day we started from the *rendezvous*. We therefore congratulated ourselves on having saved a week by commencing operations on the Thursday previous.

At noon we were in lat. 52° 11'; long. 12° 40' 2" W, eighty-nine miles distant from the telegraph station at Valentia. The water was shallow, so that there was no difficulty in paying out the wire almost without any loss by slack; and all looked upon the undertaking as virtually accomplished.

At about one o'clock in the evening the second change from the upper deck coil to that upon the orlop deck was safely effected; and shortly after the

---

[1] The amount of slack paid out had already been almost ruinous. Luckily its continuance was not necessary, or we could scarcely have reached Ireland with the cable on board.

vessels exchanged signals that they were in 200 fathoms water.

As night advanced the speed of the ship was reduced, as it was known that we were only a short distance from the land, and there would be no advantage in making it before daylight in the morning. At about twelve o'clock, however, the Skelligs Light was seen in the distance; and the *Valorous* steamed on ahead to lead us in to the coast, firing rockets at intervals to direct us, which were answered by us from the *Agamemnon*, though —according to Mr. Moriarty, the master's, wish— the ship, disregarding the *Valorous*, kept her own course, which proved to be the right one in the end.

By daylight on the morning of Thursday, the 5th, the bold and rocky mountains which entirely surround the wild and picturesque neighbourhood of Valentia rose right before us at a few miles distance. Never, probably, was the sight of land more welcome, as it brought to a successful termination one of the greatest—but at the same time most difficult—schemes which was ever undertaken. Had it been the dullest and most melancholy swamp on the face of the earth that lay before us, we should have found it a pleasant prospect; but, as the sun rose behind the estuary of Dingle Bay, tingeing with a deep soft purple the lofty summits of the steep mountains which surround its shores, and illuminating the masses of morning vapour which hung upon them, it was a scene which might vie in beauty with anything that could be produced by the most florid imagination of an artist.

No one on shore was apparently conscious of our approach, so the *Valorous* went ahead to the mouth of the harbour and fired a gun. Both ships made straight for Doulas Bay—the *Agamemnon* steaming into the harbour with a feeling that she had done something—and about 6 a.m. came to anchor at the side of Beginish Island, opposite to Valentia.

As soon as the inhabitants became aware of our approach, there was a general desertion of the place, and hundreds of boats crowded round us, their passengers in the greatest state of excitement to hear all about our voyage. The Knight of Kerry was absent in Dingle, but a messenger was immediately despatched for him, and he soon arrived in Her Majesty's gunboat *Shamrock*.

Soon after our arrival a signal was received from the *Niagara* that they were preparing to land, having paid out 1,030 nautical miles of cable, while the *Agamemnon* had accomplished her portion of the distance with an expenditure of 1,020 miles, making the total length of the wire submerged 2,050 geographical miles.

Immediately after the ships cast anchor the paddlebox boats of the *Valorous* were got ready, and two miles of cable coiled away in them, for

H.M.S. *AGAMEMNON* COMPLETING THE FIRST ATLANTIC CABLE.

the purpose of landing the end. But it was late in the afternoon before the procession of boats left the ship, under a salute of three rounds of small arms from the detachment of marines on board the *Agamemnon*, under the command of Lieutenant Morris.

The progress of the end to the shore was very slow, in consequence of the stiff wind which blew at the time; but at about 3 p.m. the end was safely brought on shore at Knight's Town, Valentia, by Mr. Bright, to whose exertions the success of the undertaking is attributable. Mr. Bright was accompanied by Mr. Canning and the Knight of Kerry. The end was immediately laid in the trench which had been dug to receive it; while a royal salute, making the neighbouring rocks and mountains reverberate, announced that the communication between the Old and the New World had been completed.

The cable was taken into the electrical room by Mr. Whitehouse, and attached to a galvanometer, and the first message was received through the entire length now lying on the bed of the sea.

Too much praise cannot be bestowed upon both the officers and men of the *Agamemnon* for the hearty way in which they have assisted in the arduous and difficult service they have been engaged in; and the admirable manner in which the ship was navigated by Mr. Moriarty materially reduced the difficulty of the Company's operations.

It will, in all probability, be nearly a fortnight before the instruments are connected at the two termini for the transmission of regular messages.

It is unnecessary here to expatiate upon the magnitude of the undertaking which has been just completed, or upon the great political and social results which are likely to accrue from it; but there can be but one feeling of universal admiration for the courage and perseverance which have been displayed by Mr. Bright,[1] and those who acted under his orders, in encountering the manifold difficulties which arose on their path at every step."[2]

In contradistinction to the heavy seas and difficulties the *Agamemnon* had to contend with, her

---

[1] In the Institution of Civil Engineers' obituary notice of Charles Bright the following lines are of some interest in this connection:—"The enormous amount of energy and resources required for the organization and fitting out of such an expedition in those early days can only with difficulty be comprehended. The details of such an undertaking are indeed massive, and reflect the very highest credit on the abilities of the late Sir Charles Bright, who (on this occasion, as on others) showed himself to be a man of extraordinary energy and power, and endowed with perseverance under difficulties—qualities which enabled him to bring this never-to-be-forgotten undertaking to a successful issue."

[2] *The Times*, Wednesday, August 11, 1858.

consort, the *Niagara*, experienced very quiet weather; and her part of the work was comparatively uneventful—with the exception of a fault near the bottom of the wardroom coil.

This was detected during the operations on the night of August 2nd, but was removed before it was paid out into the sea at a depth of two

U.S.N.S. *NIAGARA* COMPLETING THE CABLE AT THE AMERICAN END.

miles. About four o'clock the next morning the continuity and insulation was accordingly restored, and, says Mullaly, "all was going on as if nothing had occurred to disturb the confidence we felt in the success of the expedition."

A little later the same chronicler remarks:—

Confidence is growing stronger, and there is considerable speculation as to the time we shall reach Newfoundland. The pilot who is to bring us into Trinity Bay is now in great repute, and is becoming a more important personage every day. His opinion is solicited in regard to the weather, as he is supposed to know something about it in these latitudes ; and he is particularly catechized on the navigation of the bay and the formation and character of the coast. We are really beginning to have strong hopes that his services will be called into

LANDING THE AMERICAN END.

requisition, and that in the course of a few days more we will be in sight of land.

Again, when nearing the end, Mullaly describes in stirring language the various icebergs—some a hundred feet high—which they met with, whilst dilating also on their castle-like forms, etc., and the effective appearance of the sun's rays shining on them. Mullaly further gives a graphic picture of the effects of mirage here.

Shortly after entering Trinity Bay, Newfoundland, the *Niagara* was met by H.M.S. *Porcupine* (Captain Otter), which had been sent out from England at the very beginning of the 1858 ex-

NEWFOUNDLAND TELEGRAPH STATION, 1858.

pedition to await her arrival and render any assistance which might be required.

The *Niagara* anchored about 1 a.m. on August 5th, having completed her work; and during the forenoon of that day the cable was landed in a little bay, Bull Arm,[1] at the head of Trinity Bay,

[1] This spot had been selected on account of its seclusion

when they "received very strong currents of electricity through the whole cable from the other side of the Atlantic." [1]

The telegraph house at the Newfoundland end was some two miles from the beach, and connected to the cable by a land line.

SECTION XI

The Celebration.

On landing at Valentia, Charles Bright at once sent the following welcome, though somewhat startling, message to his Board, which was at once passed on to the Press :—

VALENTIA, *Aug.* 5*th*.

Charles Bright, to the Directors of the
Atlantic Telegraph Company.

The *Agamemnon* has arrived at Valentia, and we are about to land the end of the cable.

The *Niagara* is in Trinity Bay, Newfoundland. There are good signals between the ships.

We reached the rendezvous on the night of the 28th, and the splice with the *Niagara* cable was made on board the *Agamemnon* the following morning.

By noon on the 30th, 265 nautical miles were laid between the ships ; on the 31st, 540 ; on the 1st August, 884 ; on the 2nd, 1,256 ; on the 4th, 1,854 ; on anchoring at six in the morning, in Doulas Bay, 2,022.

---

from prevailing winds, and owing to the shelter it afforded from drifting icebergs.

[1] Engineer's log, U.S.N.S. *Niagara.*

THE ATLANTIC CABLE

The speed of the *Niagara* during the whole time has been nearly the same as ours, the length of cable paid out from the two ships being generally within ten miles of each other.

With the exception of yesterday, the weather has been very unfavourable.[1]

In the afternoon of Thursday, the 5th August,

LANDING THE IRISH END OF THE CABLE.

— as already described in *The Times* report —

[1] *The Times*, 2nd edition, August 5th, 1858.

Some days later Charles Bright sent in his official report, setting forth fully the main features of the expedition. Here the maximum depth was shown to be 2,400 fathoms—nearly $2\frac{1}{2}$ statute miles—and the average slack paid out somewhere about 17 per cent. This report is given in full in the Appendices to the present volume.

319

SIR CHARLES TILSTON BRIGHT

Charles Bright and his staff brought to shore the end of the cable, at White Strand Bay, near Knight's Town, Valentia,[1] in the boats of the *Valorous*, welcomed by the united cheers of the small crowd assembled.

As soon as his work was completed, Charles Bright sent his wife a telegram couched in these words : "Atlantic cable laid. Signals received both ways."[2]

All England applauded the triumph of such

---

[1] Partly by the desire of the Knight of Kerry, and partly owing to the local importance of the place, it had been decided ultimately that Knight's Town was to be the main station. Communication was, however, temporarily established by the main cable being laid by boats from the *Agamemnon* to the newly-selected terminus. A branch cable (shore-end type) was then laid across the harbour, between Ballycarberry and Knight's Town. A few days later this was underrun out to the buoyed shore end from Ballycarberry of the year before, where it was cut, and a splice effected between the seaward side and the heavy shore end.

[2] The worthy vicar of Harrow (the Rev. Edward Monro), an old friend of the family, hearing of this at the local railway station, was up at the Cedars in time to see the telegram arrive. Thus, arrayed in his usual college cap and gown, Mr. Monro was the first to shower his congratulations on Charles Bright's young wife.

Another friend soon on the spot was Mr. Arthur Noverre, who acted as physician to the family for many years.

undaunted perseverance, and the engineering and nautical skill displayed in this victory over the elements. The Atlantic Telegraph had been justly characterised by Professor Morse, as the "great feat of the century," and this was re-echoed by all the Press on its realisation. The following extract from the leading article of *The Times*, the day after completion, is an example of the comments upon the achievement :—

Mr. Bright, having landed the end of the Atlantic cable at Valentia, has brought to a successful termination his anxious and difficult task of linking the Old World with the New, thereby annihilating space. Since the discovery of Columbus, nothing has been done in any degree comparable to the vast enlargement which has thus been given to the sphere of human activity.[1]

The rejoicing in America, both in public and private, knew no bounds. The astounding news of the success of this unparalleled enterprise, after such combats with storm and sea, "created universal enthusiasm, exaltation, and joy, such as was, perhaps, never before produced by any event, not even the discovery of the Western Hemisphere. Many had predicted its failure; some from ignorance, others simply because they were anti-progressives by nature. Philanthropists everywhere hailed it as the greatest event of modern times,

[1] *The Times*, August 6th, 1858. For the rest of this article see the Appendices at end of volume.

heralding the good time coming of universal peace and brotherhood."

In Newfoundland, Mr. Field, with Captain Hudson of the *Niagara*, Captain Dayman, of H.M.S. *Gorgon*, and Commander H. C. Otter of H.M.S. *Porcupine*, together with Mr. Bright's assistant engineers, Messrs. Everett and Woodhouse, and the electricians, Messrs. De Sauty and Laws, received the heartiest congratulations and welcome from the Governor and Legislative Council of the Colony. Whilst acknowledging these congratulations, Mr. Field remarked, " We have had many difficulties to surmount, many discouragements to bear, and some enemies to overcome whose very opposition has stimulated us to greater exertion."

It was a curious coincidence that the cable was successfully completed to Valentia on the same day, in 1858, on which the shore end had been landed the year before.[1]

[1] Moreover, it was exactly one hundred and eleven years since Dr. (later Sir William) Watson had astonished the scientific world by sending an electric current through a wire two miles long, using the earth as a return circuit.

It is also worthy of note that the first feat of telegraphing was executed by order of *King Agamemnon* to his queen, announcing the fall of Troy, 1084 years before the birth of Christ, and that the great feat which we have narrated was carried out by the great *ship Agamemnon*, as we have here shown.

THE ATLANTIC CABLE

Charles Bright, with Messrs. Canning and Clifford, and the rest of the staff, as well as Professor Thomson,[1] and the electricians, were absolutely exhausted with the incessant watching and almost unbearable anxiety attending their arduous travail. Valentia proved a haven of rest indeed for these "toilers of the deep"—completely knocked up with their experiences on the Atlantic, not to mention their previous trials and disappointments.

They had really out-vied Columbus; for while he brought a vast continent within months of dangerous sea voyage, this ardent band of scientific workers had taken it within instantaneous talking distance.

Then came a series of banquets, which had to be gone through with.

Soon after his duties at Valentia were over, Bright made his way to Dublin. Here he was entertained by the Lord Mayor and civic authori-

[1] As for the Professor, in Charles Bright's subsequent words: "He was a thorough comrade, good all round, and would have taken his 'turn at the wheel' (of the brake) if others had broken down. He was also a good partner at whist when work wasn't on; though sometimes, when momentarily immersed in cogibundity of cogitation, by scientific abstraction, he would look up from his cards and ask, 'Wha played what?'"

ties of that capital on Wednesday, September 1st.
On this occasion Cardinal Wiseman, who was
present, made an eloquent speech; and the following account of the proceedings (from the *Morning
Post*) may be suitably quoted :—

The banquet given on Wednesday, the 1st, by the Lord
Mayor of Dublin, to Mr. C. T. Bright, engineer-in-chief to
the Atlantic Telegraph Company, was a great success.
The assemblage embraced the highest names in the metropolis—civil, military, and official. Cardinal Wiseman was
present in full cardinalite costume. The usual toasts were
ven, and received with all honours.

The Lord Mayor, in proposing the toast of the evening,
" The health of Mr. Bright," dwelt with much eloquence
on the achievements of science, and paid a marked and
merited compliment to the genius and perseverance which,
in the face of discouragement from the scientific world, had
succeeded in bringing about the accomplishment of the
great undertaking of the laying of the Atlantic telegraph.
His lordship's speech was most eloquent, and highly com-
plimentary to the distinguished guest, Mr. C. T. Bright.

Mr. Bright rose, amidst loud cheers, to respond. He
thanked the assemblage for their hearty welcome, and
said he was deeply sensible of the honour of having his
name associated with the great work of the Atlantic
Telegraph. He next commented upon the value of this
means of communication for the prevention of misunder-
standing between the Governments of the Great Powers,
and then referred to the services of the gentlemen who had
been associated with him in laying the cable, with whom
he shared the honours done him that night. (Mr. Bright
was warmly cheered throughout his eloquent speech.)

Mr. Bright then proposed the health of Mr. Cyrus Field, acknowledging in warm terms the services of this gentleman in the great project.

His Eminence the Cardinal descanted in glowing terms on the new achievement of science, brought to a successful issue under the able superintendence of Mr. Bright. He warmly eulogised that gentleman's modest appreciation of his services to the world of commerce and to international communication in general ; and, after paying a compliment to the Lord Mayor for his good taste in thus inaugurating in the British dominions the first public appreciation of the great work just accomplished, he proposed the health of his lordship.

The Lord Mayor returned thanks.

The health of Cardinal Wiseman was next proposed, and his eminence was again most happy in his reply.

In the subsequent toasts the railway interests of Ireland in connection with the Atlantic telegraph were eloquently responded to by Sir Edward McDonnell and Mr. Ennis M.P.

Referring next day, in a letter to his wife, to these proceedings, Bright said :—

The Cardinal came in most tremendous costume, just like Kean in *Henry VIII.*, with a large jewelled cross round his neck, and an immense sparkling ring of office on his white hand, which contrasted strongly with his red face and dress. However, I found him a very pleasant man, full of scientific knowledge and interest in the Atlantic line. He pressed me to come to see him in London.

I hope you thought my speech a good one ! I wanted

to have a public opportunity of shaming the "Yankees" by proposing Cyrus Field's health.[1]

In wishing to put on record in a permanent and suitable form their appreciation for the work carried out by Charles Bright and his staff, the Directors of the Atlantic Telegraph Company caused an illuminated testimonial to be drawn up, a *fac-simile* of which is here reproduced.[2]

Charles Bright was honoured with knighthood within a few days of his landing.

As this was considered a special occasion—apart from ordinary periodic honours—and as the Queen was at that time on her famous and important visit to the Empress of the French at Cherbourg, it was arranged that the ceremony should be performed there and then by His Excellency the Lord Lieutenant of Ireland (the Earl of Eglinton) in Her Majesty's name.[3]

---

[1] Our friends "on the other side" had omitted to even allude to Charles Bright amidst all their enthusiasm and rejoicing, though the work had been carried out by him as the engineer! Indeed, altogether, with an overdose of patriotism, they were rather forgetful of their "cousins" and of the actual facts of the case.

[2] The arranging of this (by Mr. Cyrus Field) was not, however, quite happy, for there is no question that Mr. (now Sir Samuel) Canning was Charles Bright's chief assistant, and that Mr. Everett—Mr. Field's fellow countryman—occupied a subordinate.

[3] The following spring Charles Bright was duly "presented" at Court in connection with his knighthood.

At a Meeting of the Board of Directors of the Atlantic Telegraph Company held at 22 Old Broad Street in the City of London on Tuesday the 10th day of August 1858

It was Resolved

That the Thanks of this Board be given to Mr Charles Gideon Bright and Mr W E Everett as also to Mr Samuel Canning Mr William Henry Woodhouse Mr Henry Clifford and the other Officers of the Company who have acted on Board the Agamemnon and Niagara for the Services they have severally conferred on this Company and for the Great Zeal and Energy they have displayed under difficult and dangerous circumstances in carrying out their Important Duties during the submerging of the Atlantic Cable

The Board further desire to Congratulate them upon the part they have taken in bringing a work of so truly national a character to a Successful Termination

Given under the Common Seal of the Company the 10th day of August 1858

THE ATLANTIC CABLE

With reference to this, Bright wrote to his wife the day before :

The Lord Lieutenant having expressed a wish to see me, I had an interview with him this morning. He intimated his desire to confer upon me, on behalf of Her Majesty, the honour of Knighthood, which ceremony is to be performed to-morrow, after which I dine with him to meet a large party of the noble folks of the land, and then I shall be glad to get home and have a little quiet with " Lady Bright."

Bright was but twenty-six years of age at the time, being the youngest man who had received the distinction for generations past—and we know of no similar instance occurring since. It was the first title conferred on the telegraphic profession, and remained so for many years.

Captain George Preedy, the gallant commander of H.M.S. *Agamemnon*, was made a C.B., as was also Captain W. C. Aldham, of H.M.S. *Valorous*. Other officers received promotion.

With Professor Thomson and other colleagues Sir Charles was right royally entertained in Dublin, Killarney, and elsewhere, the Lord Lieutenant taking a prominent part in the celebrations.[1] Indeed, in Ireland generally, where he

[1] It was just previous to one of these at the Vice-Regal Lodge that opportunity was taken to perform the ceremony of "knighting" Bright. At the dinner afterwards he sat next to the then Duchess of Manchester;

had been previously known for years as the able engineer of the " Magnetic" Company — whose wires he had extended throughout the length and breadth of the Emerald Isle—the warm greetings were unbounded.

A few days later, on the occasion of the grand banquet given in his honour at Killarney by the nobility and gentry of Kerry, His Excellency the Lord Lieutenant, after some prefatory remarks, thus referred to him and the cable[1]:—

When we consider the extraordinary undertaking that has been accomplished within the last few weeks ; when we consider that a cable of about 2,000 miles has been extended beneath the ocean—a length which, if multiplied ten times, would reach our farthest colonies and nearly surround the earth ; when we consider it is stretched along the bed of shingles and shells, which appeared destined for it as a foundation by Providence, and stretching from the points which human enterprises would look to ; and when we consider the great results that will flow from the enterprise, we are at a loss here how sufficiently to admire the genius and energy of those who planned it, or how to be sufficiently thankful to the Almighty for having delegated such a power to the human race, for whose benefit it is to be put in force. (Cheers.) And let us look at the career which this tele-graph has passed since it was first discovered. At first it was rapidly laid over the land, uniting states, com-

who reminded him that one of his ancestors had married Lady Lucy Montague—one of the Duke's family—of previous days.   [1] *Daily News*, August 20th, 1858.

munities and countries, extending over hills and valleys, roads and railways; but the sea appeared to present an impenetrable barrier. It could not stop here, however; sub-marine telegraphy was but a question of time, and the first enterprise by which it was introduced was in connexion with an old foe—and at present our best friend —Imperial France. (Hear, hear.) The next attempt which was successful was the junction of England and our island, and which was, I believe, carried out by the same distinguished engineer, whose name is now in the mouth of every man. (Hear, hear.) Other submarine attempts followed: the telegraph paused before the great Atlantic, like another Alexander, weeping as if it had no more worlds to conquer; but it has found another world, and it has gained it—not bringing strife or conquest, but carrying with it peace and goodwill. (Applause.) I feel I should be wanting if I did not allude in terms of admiration to the genius and skill of the engineer, Sir Charles Bright, who has carried out this enterprise, and to the zeal and courage of those who brought it to a successful termination. (Applause.) It is not necessary, I am certain, to call attention to the diligence and attention shown by the crew of the *Agamemnon*—(cheers)—because I am sure there is no one here who has not read the description of the voyage in the newspapers. The zeal and enterprise were only to be equalled by the skill with which it was carried out. I believe there was only a difference of twelve miles between the two ends of the cable when it came to the shore. There are some questions with regard to the date at which the work was carried out, to which I wish to call attention. It was on the 5th August, 1857, that this enterprise was first commenced under the auspices of my distinguished predecessor, who I wish was here now to rejoice in its success—I mean only in a private capacity. (Cheers and laughter.) It was on the 5th August, 1858, it was completed, and it was on the 5th August, more than 300 years ago, that Columbus left the shores of Spain to proceed on his ever-memorable voyage to America. It was on the 5th of August, 1583, that Sir Hugh Gilbert, a worthy countryman of Raleigh and Drake, steered his good ship the *Squirrel* to the shores of Newfoundland, and first unfurled the flag of England in the very bay where this triumph has now taken place—(applause)—and it was on the same 5th of August that your sovereign was received by her imperial friend amidst the fortifications of Cherbourg, and thereby put an end to the ridiculous nonsense about strife and dissension. (Applause.) Let the 5th August be a day ever memorable among nations. Let it be, if I may so term it, the birthday of England. (Applause.) Among the many points which must have given every one satisfaction, was the manner in which this great success was received in America. (Hear.) There appears to have been but one feeling of rejoicing predominant amongst them ; and I cannot but think that that was not only owing to their commercial enterprise—which they shared along with us— but also, I trust, more to the feelings of consanguinity and affection which I am sure we share, though occasionally disturbed by international disputes, and by differences caused by misrepresentations or hastiness. It must still burn as brightly in their breasts as in ours. (Applause.) I trust that, not only with our friends across the Atlantic, but with every civilised nation, this great triumph of science will prove the harbinger of peace, goodwill, and friendship ; and that England and America will not verify the first line of the stanza—

THE ATLANTIC CABLE

"Lands intersected by a narrow firth
Abhor each other,"—

but that they will, by mutual intercourse, arrive at the last line of that stanza, and, "like kindred drops, be mingled into one." (Warm applause.)

After the various functions in Ireland celebrating the laying of the cable had been exhausted, Bright was glad to have the opportunity of returning to his family at Harrow Weald, for the first time since the successful completion of the work.

Section XII

The Working of the Line

As previously shown, two descriptions of instruments were used on board the ships for testing and working through whilst laying the cable. These were the detector of Mr. Whitehouse and Professor Thomson's reflecting apparatus.

The process of testing consisted in sending from one to the other vessel alternately, during a period of ten minutes, first a reversal every minute for five minutes, and then a current in one direction for five minutes. The results were observed and recorded on board both ships. There was also a special signal for each ten miles of cable paid out between the vessels.

When the splice was made on the 29th July, $72°$ deflection were obtained on the *Agamemnon* from seventy-five cells of a sawdust Daniell's battery on board the *Niagara* which had given 83° on entry.

On arrival at Valentia at 6.30 a.m. on the 5th of August, the deflection on the same instruments (detector and marine galvanometer being both in circuit as before) was 68°; while the sending battery power on the *Niagara* had fallen off at entry to $62\frac{1}{2}$° through the marine galvanometer on board that vessel.[1]

The figures quoted show that, taking into account the certain diminution in electro-motive force of the " sawdust " battery employed, the cable had considerably improved by submersion, the insulation being even greater than that recorded before laying, when the cable was reported as perfect.

When Charles Bright and his staff had accomplished their part of the undertaking on the 5th August, the cable was handed over to Mr. Whitehouse, the electrician of the Company, and his assistants. It was then reported to be in perfect condition.

Mr. Whitehouse, however, after taking charge of the line, found difficulty in working it with his special induction apparatus,[2] but appears to have

[1] For full particulars see " Contributions for a History of the Atlantic Cable," the *Electrician*, 1863.

[2] Besides being fully described in the pages of the *Engineer* at the time, some of Mr. Whitehouse's apparatus

made no report to the Board for a week.   No information arrived at head-quarters except some telegrams stating that signals were highly satisfactory, and that the adjustment of instruments was progressing.[1]

More than a week passed, during which Mr. Whitehouse continued his ineffectual efforts to work with the induction apparatus; and then Professor Thomson's reflecting galvanometer, that had worked so well during the voyage, was again inserted, with ordinary Daniell cells, in the circuit.

In this way communication was resumed, the first clear message being received from Newfoundland on the 13th of August, 1858, and on the 16th the following message was got through from the directors in England to the directors in America :[2]—

may now be seen at Messrs. Elliott, Brothers, the famous instrument makers of St. Martin's Lane, London.

[1] *The Transatlantic Submarine Telegraph*, p. 33, by George Saward, Secretary to the Atlantic Telegraph Company.

[2] There had been a considerable delay in getting the apparatus ready at Newfoundland; and unfortunately they adhered to alternating electro-magnetic apparatus there in conjunction with a relay.   The result was that supreme difficulty was experienced throughout in working the line this way.   On the other hand, at Valentia they once reported: "We are now receiving from Newfoundland accurately, at the rate of 100 words per hour."   In-

" Europe and America are united by telegraph. 'Glory to God in the highest, on earth peace, good-will toward men.'" [1]

Then followed :—

From Her Majesty the Queen of Great Britain to His Excellency the President of the United States.

The Queen desires to congratulate the President upon the successful completion of this great international work, in which the Queen has taken the greatest interest.

The Queen is convinced that the President will join with her in fervently hoping that the electric cable, which now already connects Great Britain with the United States, will prove an additional link between the two nations, whose friendship is founded upon their common interest and reciprocal esteem.

The Queen has much pleasure in thus directly communicating with the President, and in renewing to him her best wishes for the prosperity of the United States.

---

deed, nearly all the really successful working was effected by the Thomson " marine galvanometer," at a speed up to 5 words per minute, as compared with 1·75 per minute with the other apparatus.

[1] With reference to this and some of the following cablegrams, Sir D. Brewster wrote (in the *Edinburgh Review*) at the time: " It is impossible to read, without emotion, these messages which breathed—from the earliest to the latest—the ardent wish that peace and good-will should reign between hitherto unfriendly nations, born of the same blood, speaking the same tongue, and rejoicing in the same faith."

THE ATLANTIC CABLE

This message was shortly afterwards responded to as follows :—

WASHINGTON CITY.

The President of the United States to Her Majesty Victoria, Queen of Great Britain.

The President cordially reciprocates the congratulations of Her Majesty the Queen on the success of the great international enterprise accomplished by the skill, science, and indomitable energy of the two countries.

It is a triumph more glorious, because far more useful to mankind, than was ever won by a conqueror on the field of battle.

May the Atlantic Telegraph, under the blessing of Heaven, prove to be a bond of perpetual peace and friendship between the kindred nations, and an instrument destined by Divine Providence to diffuse religion, civilisation, liberty, and law throughout the world !

In this view will not all the nations of Christendom spontaneously unite in the declaration that it shall be for ever neutral ; and that its communications shall be held sacred in passing to the place of their destination, even in the midst of hostilities ?

JAMES BUCHANAN.

Throughout the United States the arrival of the Queen's message was the signal for a fresh outburst of popular enthusiasm.[1]

---

[1] Whoever shall write the history of popular enthusiasm must give a large space to the way in which the advent of Atlantic telegraphy was received in the United States. Never did the tidings of any great achievement—whether of peace or war—more truly electrify a nation. In New

SIR CHARLES TILSTON BRIGHT

Says Field:

The next morning, August 17th, the city of New York was awakened by the thunder of artillery. A hundred guns were fired in the City Hall Park at daybreak, and the salute was repeated at noon. At this hour flags were flying from all the public buildings, and the bells of the principal churches began to ring, as Christmas bells signal the birthday of one who came to bring peace and goodwill to men—chimes that, it was fondly hoped, might usher in, as they should, a new era.

> "Ring out the old, ring in the new,
> Ring out the false, ring in the true."

That night the city was illuminated. Never had it seen so brilliant a spectacle. Such was the blaze of light around the City Hall that the cupola caught fire and was consumed, and the Hall itself narrowly escaped destruction. But one night did not exhaust the public enthusiasm, for the following evening witnessed one of those displays for which New York surpasses all the cities of the world—a fireman's torchlight procession. Moreover, several wagon-loads (each containing about twelve miles) of the cable left on board the *Niagara* were drawn through the principal streets of the city.

Similar demonstrations took place in other parts of the United States. From the Atlantic to the valley of the

---

York the news was received at first with incredulity. No doubt the impression was greater, because it took every one completely by surprise. This undertaking had been looked upon as hopeless. Its projectors had shared the usual lot of those who conceive vast designs and venture on great enterprises, and their labours had been watched with mixed feelings of derision and pity.

THE ATLANTIC CABLE

Mississippi, and to the Gulf of Mexico, in every city was heard the firing of guns and the ringing of bells. Nothing seemed too extravagant to give expression to the popular rejoicing.

In an able article on the subject, Sir David Brewster, F.R.S., wrote [1] :—

Our countrymen in the British provinces were not less enthusiastic than their brethren in the States. At St. John's, New Brunswick, the general joy was expressed by the firing of guns and the display of fireworks; and the city was in a perfect blaze from the illumination of the public offices, warehouses, and private buildings. At Halifax, Quebec, Toronto, Chatham, Hamilton, and Montreal there was the same excitement. The firing of guns, the display of flags, fireworks, illuminations, and processions were everywhere the voluntary expressions of public feeling.

That men in official positions, and the mercantile classes, should have celebrated the success of the Atlantic Telegraph is hardly a matter of surprise; but we were not prepared to expect that the masses of the American population in our own Colonies, as well as in the States, should display so much sympathy and exultation.

The English press were warm in their recognition of those to whom the nation were "indebted for bringing into action the greatest invention of the age," [2] and expressed their full belief that "the effect of bringing the Three Kingdoms and the United States into instantaneous communication

[1] *North British Review*, November, 1858, p. 543.
[2] *The Times*, August 6, 1858.

with each other, will be to render hostilities between the two nations almost impossible for the future."

" More was done yesterday for the consolidation of our Empire than the wisdom of our statesmen, the liberality of our Legislature, or the loyalty of our Colonists could ever have effected."

The sermons preached on the subject, both in England and America, were literally without number. Enough found their way into print to fill large volumes. Never had an event more deeply touched the spirit of religious enthusiasm.

With further reference to the active life of the cable, the following communications have some interest :—

Three long congratulatory messages were transmitted : one on the 18th August, from Mr. Peter Cooper, President of the New York, Newfoundland, and London Telegraph Company, to the Directors of the Atlantic Telegraph Company ; another, from the Mayor of New York to the Lord Mayor of London, his reply in acknowledgment following.

Two of the great Cunard mail steamers, the *Europa* and *Arabia*, came into collision on the 14th August, while on their outward and homeward voyages. Neither the news nor the injured vessels could reach those concerned on both sides

THE ATLANTIC CABLE

of the Atlantic for some days; but as soon as it became known in New York, a message was sent by the cable :—

"*Arabia* in collision with *Europa*, Cape Race, Saturday. *Arabia* on her way. Head slightly

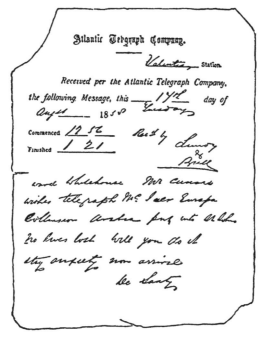

injured. *Europa* lost bowsprit, cutwater stem sprung. Will remain in St. John's ten days from 16th. *Persia* calls at St. John's for mails and passengers. No loss of life or limb."

This first public *news* message showed the relief given by speedy knowledge in dispelling doubt and fear. A New York Stock Exchange gentleman remarked at the time : "I was in Liverpool expecting friends by the *Europa*. Any delay in the arrival of the ship would have caused great anxiety. But one morning, on going down to the Exchange, we saw posted up this despatch, received the night before by the Atlantic Telegraph. All then said, if the cable never did anything more, it had amply repaid its cost."

Subsequently, messages giving the news on both Continents were transmitted, and published daily. An early one to America read as follows :—

"*North American* with Canadian, and *Asia* with direct Boston, mails, leave Liverpool, and *Fulton*, Southampton, Saturday next. To-day's morning papers have long, interesting reports from Bright."

"Indian News, *Virago* arrived at Liverpool to-day. Bombay dates 19th July, mutiny being rapidly quelled."

The news of peace with China shortly followed. Amongst others, on 27th August, a despatch that —looking at the variety of information conveyed —was remarkable for the conciseness displayed by its sender, the Secretary of the Atlantic Company :—

"To Associated Press, New York. News for America by Atlantic Cable : Emperor of France returned to Paris, Saturday. King of Prussia too ill to visit Queen Victoria. Her Majesty returns to England 30th August. St. Petersburg, 21st August : Settlement of Chinese Question ; Chinese Empire opened to trade ; Christian religion allowed, foreign diplomatic agents admitted ; indemnity to England and France. Alexandria, August 9th, the *Madras* arrived at Suez 7th inst. Dates Bombay to the 19th, Aden, 31st. Gwalior insurgent army broken up. All India becoming tranquil."

The above was published in the American papers the same day, many items of news from the States and Canada also appearing in the English press.

Further, as exemplifying the aid the cable afforded to our Government, we may mention two messages sent from our Commander-in-Chief, at the Horse Guards, on the 31st August—owing to the quelling of the Indian Mutiny — cancelling orders sent by mail to Canada, thus :—

The first, to General Trollope, Halifax, ran as follows : " The 62nd Regiment is not to return to England." The other, to the officer in command at Montreal, ran thus : " The 39th Regiment is not to return to England."

From £50,000 to £60,000 was estimated by the authorities to have been saved in the unnecessary transportation of the troops by these two cable communications, which were delivered the same day that they were sent.

But the insulation of the precious wire had,

unhappily, been giving way, and the diminished flashes of light proved to be only the flickering of the flame that was soon to be extinguished in the eternal darkness of the waters.

After a period of confused signals, the line ultimately breathed its last on October 20th, after 732 messages[1] in all had been conveyed during a period of three months. The last word which the line uttered — and which may be said to have come beyond the sea — was " Forward !" The very day that a whole city rose up to do honour to the Atlantic Telegraph—when the roar of guns, the chiming of bells in the sacred spires, and the shouts of joy throughout the land might be heard o'er hill and dale, and when London also was about to do it honour—the throbs of this almost living thing were becoming visibly weaker, and fears began to prevail that it would shortly sleep for ever silent in its ocean grave.

The line had been subject to frequent interruption throughout. The wonder is that it did so much, when we consider the lack of experience at that period in the manufacture of deep sea cables, the short time allowed, and the treatment the line received after being laid.[2]

---

[1] Of these a large number were merely between the operators themselves, respecting the working arrangements, which, whilst of direct service to them, were of no interest to the general public.

[2] It is extremely doubtful whether any cable, even of

THE ATLANTIC CABLE

An unusually violent lightning storm occurred at Newfoundland shortly after the cable had been laid. This was spoken of as a possible part-cause of the gradual failure of the line : also a supposed "factory fault," masked by the tar in the hemp.

There were, however, those who hinted at foul play. It was certainly singular that the cable should continue to work for several weeks and only show definite signs of sickness on the very day of the celebration in New York !

When all the efforts of the electricians failed to draw more than a few faint whispers—a dying gasp from the depths of the sea—there ensued, in the public mind, a feeling of profound discouragement. And then, as regards those officially concerned in the enterprise. What a bitter disappointment ! Imagine Charles Bright's state of mind after all he had gone through, and after he had ultimately successfully accomplished his part of the undertaking ! In all the experience of life there are no sadder moments than those

the present day, would long stand a trial with currents so generated and of such intensity.

In his work on the "Electric Telegraph," the late Mr. Robert Sabine said : "At the date of the first Atlantic cable the engineering department was far ahead of the electrical. The cable was successfully laid—mechanically good, but electrically bad."

in which, after years of anxious toil, striving for a great object, and after a glorious triumph, the achievement that seemed complete becomes a wreck.

Still, young Bright had the satisfaction of knowing that he had (1) demonstrated the possibility of laying over 2,000 miles [1] of cable in one continuous length across the Atlantic Ocean at depths of two to three miles [2]; and (2) that by means of an electric current, distinct and regular signals could be transmitted and received through an insulated conductor, even when at such a depth beneath the sea, across this vast distance.[3]

---

[1] This was a length six times greater than had ever been previously laid, and at an average depth far in excess of anything that had been done before.

[2] He had also proved, amongst other things, that a ship could be hove to in deep water with a cable hanging on without the latter breaking.

[3] In his Presidential Address to the Institution of Electrical Engineers, in 1889, Lord Kelvin (then Sir William Thomson) said :

The first Atlantic cable gave me the happiness and privilege of meeting and working with the late Sir Charles Bright. He was the engineer of this great undertaking—full of vigour, full of enthusiasm. We were shipmates on the *Agamemnon* on the ever memorable expedition of 1858, during which we were out of sight of land for thirty-three days. To Sir C. Bright's vigour, earnestness, and enthusiasm was due the successful laying of the cable. We must always feel deeply indebted to our late colleague as a pioneer in that great work, when other engineers would not look at it, and thought it absolutely impracticable.

THE ATLANTIC CABLE

Of course the gutta percha coverings, as then applied in almost the earliest stages of submarine work,[1] cannot in any way be compared to the continual progress made in insulating methods and materials during the many years that have since elapsed. But in 1856-57 the Atlantic cable insulation was a great advance upon that applied to the wires of previous cables : moreover, the conductor was a *strand* of copper, and much larger than anything previously adopted.

It was to be regretted that owing to the precipitate orders given by the provisional committee of the subscribers to the memorandum of association of the Company—before even the Board had been formed or Charles Bright appointed engineer—that his specification of a conductor (nearly four times larger) of about the same size and insulated with much the same amount of gutta percha as subsequently employed eight years later for the Atlantic cables of 1865 and 1866, had not been adopted.[2]

Bright's specification—could it have been adopted

---

[1] It was thought by some that the gutta percha had let the water percolate in at the seams, and also that weak joints contributed to the ultimate failure of the lime.

[2] This heavier type of core was also recommended to Government by Sir C. Bright, in 1860, for the Falmouth-Gibraltar cable, eventually used to connect Malta and Alexandria.

by the already fettered Board—would have given six times the insulation, and at the same time more than treble the conductivity. Under such conditions it is highly improbable that strong currents would have been applied for the working of the line. Unhappily Professor Morse had promulgated an opinion directly opposed to Charles Bright's practical knowledge.

Professor Morse's views ran thus :—

That large-coated wires used beneath the water or the earth are worse conductors—so far as velocity of transmission is concerned—than small ones ; and, therefore, are not so well suited as small ones for the purposes of submarine transmission of telegraphic signals.

Further :—

That by the use of comparatively small-coated wires, and of electro-magnetic induction coils for the exciting magnets, telegraphic signals can be transmitted through two thousand miles, with a speed amply sufficient for all commercial and economical purposes. [1]

A similarly incorrect theory was adopted even by Faraday (the greatest electrical scientist of the day), who, in a discussion on the proposed Atlantic Cable at the Institution of Civil Engineers, stated " that the larger the jar, or the larger the wire, the more electricity was required to charge it ; and the greater was the retardation of that electric impulse,

---

[1] Report by Professor S. F. B. Morse, LL.D., to the Provisional Committee of the Atlantic Telegraph Company.

which should be occupied in sending the charge forward," [1] thereby entirely disregarding the factor of conductor resistance.

The Company were completely misled by this and by the similar views entertained by Mr. Whitehouse. And so to a cable of comparatively small carrying power and poor insulation was set the task of withstanding electric currents of an intensity that would probably ruin any line ever laid, even now—forty years later!

Still, the cable, inadequately constructed as it was, would probably have worked—though slowly, of course—for years, had the fairly reasonable battery currents employed between the ships, and up to the successful end of the voyage, been continued in connection with Professor Thomson's very delicate reflecting apparatus. Mr. Whitehouse, however, connected his battery to fearfully intense induction coils in order to work his specially devised relay and Morse electro-magnetic recording instruments at the further end of the line. Moreover, finding difficulty in getting his appliances to act properly, he appears to have increased the power from time to time, up to nearly 500 cells— of a very potent type—during the first week of working, till the induction coils of about *five feet* long yielded a current that was estimated by the experts (who sat at a sort of coroner's inquest

---

[1] Professor Michael Faraday in *Proceedings of Inst. C.E.*, vol. xvi., p. 221.

on the unhappy cable) to amount to about 2,000 volts!

Hence, when signalling was resumed by the comparatively mild voltaic currents, actuating Professor Thomson's instrument, a fault (or faults) had been already developed, necessitating a far higher battery power than had been employed during the continuous communication between the ships whilst paying out.

The wounds opened further under the various stimulating doses; the insulation was unable to bear the electrical strain, and the circulation gradually ceased through a cable already in a state of dissolution.

The services of Mr. J. R. France were placed at the disposal of Mr. Whitehouse for testing and endeavouring to re-establish communication—but with no effect.

SECTION XIII

The Inquest

The great historical sea line having collapsed, some of the foremost of the electrical profession were called in to aid, first in determining the *nature* of the interruption, with a view to remedies if practicable, next to elicit the *cause*.

Mr. C. F. Varley,[1] the electrician of the Electric

---

[1] About this time, or shortly afterwards, Mr. Varley

Telegraph Company; Mr. E. B. Bright, the chief of the Magnetic Company; and Mr. W. T. Henley, the well-known telegraph inventor, were severally requested by the Atlantic Company to examine and report in conjunction with Sir Charles Bright and Professor Thomson.

Resistance coils and apparatus for ascertaining the position of the fault, patented by the Messrs. Bright in 1852—as referred to in Chapter III.—were employed; and indicated a serious leakage of electricity at a distance of about 300 miles from Valentia. It was obvious there was no fracture of the conductor, for excessively weak currents still came through fitfully.

According to the above location the main leak through the gutta-percha envelope was in water of a depth of about two miles. At that time means were not devised for grappling and lifting a cable from such depths. In fact, the first serious attempts to do so in an open sea-way were made in 1865, when the Atlantic cable of that year broke in mid-ocean. But the appliances then failed, and it was not until after the laying of the third cable, in 1866, that the general arrangements were sufficiently well worked out, or the grappling ropes

became electrician to the Atlantic Company in succession to Mr. Whitehouse, who had retired, whilst Professor Thomson still remained scientific adviser to the Board of Directors.

and shackles made strong enough to raise and repair the 1865 cable.

As the result of tests made independently by Charles Bright and Professor Thomson, it appeared likely that the Valentia shore end was especially faulty. Accordingly it was under-run from the catamaran raft—previously used in 1857 —for some three miles; but on being cut at the furthest point at which it was found possible to raise the cable, the fault still appeared on the seaward side. The idea of repairs had, therefore, to be abandoned, and the cable was spliced up again.

The conductors being again intact, efforts were made to renew signals, by means of a large and improved magnetic-telegraph devised by Mr. Henley, and also by curb keys recently invented by Mr. E. B. Bright,[1] by means of which currents of opposite character, and of given lengths, were transmitted so that each signalling current was followed instantly by one of opposite polarity, which neutralized, by a proportionate strength and duration, all that remained of its predecessor. The road was thus cleared for the succeeding signal, varying from the first so as to form letters. This "curb" was a modification of the apparatus devised with a similar object and patented in 1852 by Messrs. Bright, and was

---

[1] Patent No. 54 of 1858.

subsequently greatly improved by Sir Charles in 1860 and 1862.[1]

All efforts, however, proved unavailing : for signalling purposes the poor cable was defunct.

Having dealt with the nature of the interruption, we now come to the *cause*.

It is first of all abundantly clear from the station diaries kept by the electricians at Valentia and Newfoundland, and by other irrefragable evidence, that when the laying was completed, and the cable ends were handed over to them from the ships on the 5th August, all was in good working order.

Thus :—

"On the landing of the cable at Newfoundland some of them 'tasted' the current, and received a pretty strong shock, so strong that they willingly resigned the chance of repeating the experiment." On the same day, the 5th August, Mr. Field telegraphed to the New York Associated Press : "The electrical signals sent through the whole cable are *perfect*." The station diary records the same ; again, on the 8th August, the entry runs : "Good signals being received through the cable." On the 9th Mr. De Sauty, the electrician, reports : "'Receiving' good, 'recorded' signals from Valentia. *Perfectly satisfactory*."

So much for the American end. On this side it was

---

[1] Sir C. T. Bright's patents, No. 465, 20th February, 1860 ; No. 538, 27th February, 1862.

stated in the papers on 5th August, "good signals passing to and fro." Mr. Whitehouse, the chief electrician, reports on the 6th: "Electric communication is maintained perfectly." 7th: "The currents from Newfoundland are good, *giving deflections* of 60° on either side of the galvanometer, according as a positive or negative current is transmitted." On the 10th August Mr. Whitehouse telegraphed that "rate of transmission fully equals that obtained at Keyham, and the line works as well as it did before it was laid."[1]

Mr. Whitehouse, in his evidence before the Government Commission appointed to enquire into the construction of submarine cables, stated:[2] "The signals were *very strong*: they made the *relay* speak out *loud*, so that *you could hear it across the room.*[3] The battery power employed at the time at Newfoundland was seven twelve-cell sawdust batteries, *i.e.* the ordinary portable copper and zinc battery in a gutta percha case, but filled with sawdust instead of sand."

On board the ships during the submersion, only comparatively moderate charges of electricity were employed for signalling—some seventy cells of a very ordinary, and weak, form of voltaic battery. The use of these was continued at Valentia after landing, and worked the cable perfectly, though relatively slowly (as was expected) compared with overhead land wires.

[1] The preceding statements and remarks are taken from the records of the *Electrician.*
[2] *Blue Book*, 1861.   [3] *Ibid.*

THE ATLANTIC CABLE

It is true that the log of the *Agamemnon* recorded a cessation of signals and apparent earth.

August 2nd.—Throughout the greater portion of Monday morning the electrical signals from the *Niagara* had been getting gradually weaker, until they ceased altogether for three quarters of an hour ! . . . Accordingly Professor Thomson sent a message to the effect that the signals were too weak to be read : and, as if they had awaited such a signal to increase their battery power, the deflections immediately returned even stronger than they had ever been before. Towards the evening, however, they again declined in force for a short time.[1]

But on comparing the log of the *Niagara* for the same day, Monday, August 2nd, we find :—

" About noon, imperfect insulation of cable detected in sending and receiving signals from the *Agamemnon*, which continued till forty minutes past five, when all was right again. The fault was found to be in the ward-room, about sixty miles from the lower end, which was immediately cut out and taken out of the circuit."

Curiously this, the cutting out of a fault on board—an ordinary incident in cable laying—was afterwards erroneously magnified into a fault going overboard, for a want of the above comparison of the two logs.

Now there were only *time* signals between the ships, and no "speaking" arranged for. So while this fault existed—and for the hours that it was being got at, *and cut out*—on the *Niagara*, Charles

[1] Report in *The Times*.

355

Bright and his staff, with Professor Thomson and the electricians, engaged on the British moiety, were in ignorance of the cause of the cessation of signals. At the time, therefore, it was necessarily put down by them to some fault at sea which mysteriously rectified itself.

All the eminent and practical electricians examined before the Parliamentary Committee were unanimous on this point :—[1]

Mr. Cromwell Varley, F.R.S., declared his belief "that had a more moderate power been used, the cable would still have been capable of transmitting messages," and that its faulty condition was no doubt due to the employment of large induction coils.[2]

[1] "Joint Commission on Submarine Telegraph Cables." —*Ibid.*

[2] In giving extra force to the above opinion, Mr. Varley then described an experiment which he had made on the cable when asked to report by the directors of the Atlantic Company on its condition. He said : " To satisfy myself on this point I attached to the cable a piece of gutta-percha covered wire, having first made a slight incision, by a needle prick, in the gutta percha to let the water reach the conductor. The wire was then bent, so as to close up the defect. The defective wire was then placed in a jar of sea water, and the latter connected with the earth. After a few momentary signals had been sent from the five-feet induction coils into the cable—and, consequently, into the test wire—the intense current burst

Mr. J. W. Brett (a Director), stated that "the Board had clear evidence that the cable sustained injury by the use of very great power."

Mr. Glass "was persuaded that the intense currents were finally the cause of the signals ceasing."

Professor Hughes, the inventor of the well-known type-printing telegraph, and subsequently of the microphone, declared that "the cable was injured by the induction coils, and that the intense currents developed by them were strong enough to burst through gutta-percha." [1]

A member of the committee afterwards enquired whether it was the fact, that those who had the misfortune to touch the cable at the time when the current was discharged from the induction coil, received so severe a shock from it that they nearly fainted. It was admitted in reply, "that those who touched the bare wire would suffer for their carelessness, though not, if *discretion were exercised*, in grasping the gutta-percha only."

Professor Wheatstone expressed his opinion at

through the excessively minute perforation, rapidly burning a hole nearly one-tenth of an inch in diameter, afterwards increased to half an inch in length when passing the current through the faulty branch only. The burnt gutta percha then came floating up to the surface of the water, whilst the jar was one complete glow of light."

[1] "Contributions," etc., from the *Electrician*, 1863, p. 15.

the enquiry in question : " That the force of the induction coils must have been enormously greater than that of a battery of 400 elements," such as we subsequently employed at Valentia in the later signalling efforts.

Further evidence was given to the same effect by other experts ; and the Right Hon. J. Stuart Wortley, M.P., the then Chairman of the Atlantic Company, in a deputation to Lord Palmerston in March, 1862, stated that[1] " far too high charges of electricity, were forced into the conductor. It was evidently thought at that time by certain electricians, that you could not charge a cable of this sort too highly. Thus, they proceeded somewhat like the man who bores a hole with a poker in a deal board : he gets the hole, to be sure, *but the board is burnt in the operation.*"

This remark was doubtless inspired by the experiment above referred to and made by Mr. C. F. Varley, with Mr. Edward Bright, at Valentia.

Professor Thomson (now Lord Kelvin), writing in 1860, expressed the following opinion anent the use of excessive power :—

" The induction coils were superseded by Daniell's battery at Valentia after a few days' trial, though the rapidly failing line had seemed to prove them incapable of giving intelligible signals to the Newfoundland Station. Owing to the immediate introduction and continued use of an entirely new

---

[1] " Contributions " from the *Electrician* ; 1863, p. 19.

kind of receiving instrument, the mirror galvano-meter—introduced for long submarine telegraphs by the writer—at Valentia, the signals from the New-foundland coils were found sufficient during the three weeks of successful working of the cable."

" It is quite certain that, with a properly adjusted mirror galvanometer as receiving instrument at each end, twenty coils of Daniell's battery would have done the work required, and at even a higher speed if worked by a key devised for diminishing inductive embarrassment ; and the writer, with the knowledge derived from disastrous experience, has now little doubt but that, if such had been the arrangement from the beginning, if no induction coils and no battery power exceeding twenty Daniell cells had ever been applied to the cable since the landing of its ends, imperfect as it then was, *it would be now in full work day and night, with no prospect or probability of failure."* [1]

Summing up the *cause* of the catastrophe to the ill-used cable, perhaps the curtest verdict would be that (in mechanical engineering parlance) " high pressure steam had been got up in a low pressure boiler."

[1] *The Encyclopedia Britannica*, 8th Edition, 1860. Article on " The Electric Telegraph," by Professor W. Thomson, F.R.S.

SIR CHARLES TILSTON BRIGHT

The North Atlantic Route

It soon became evident that no fresh venture would take practical shape for several years. Seeing this, Sir Charles devoted himself to his other professional work, connected with the Magnetic Telegraph Company, and subsequently to the accomplishment of other undertakings of which the first line to India—superintended and laid by him for Government, in 1864, along the Persian Gulf—was the chief.

It was not, however, as it turned out, so very long before he became interested in another big Atlantic cable project.

The failure of the first line after a short period of working, and the slow rate at which messages were capable of being passed through its conductor, naturally deterred capitalists from providing the means for another cable, of such length in deep water.

But there was an alternative route between this country and America, by which the transmission of the electric current could be subdivided into four comparatively short circuits : namely — from the extreme north of Scotland, to the Faröe Islands, thence to Iceland, from there to the southern point of Greenland, and so to Labrador, or Newfoundland. Although this route looks much longer on the map, it is not

THE NORTH ATLANTIC TELEGRAPH PROJECT, 1860.

really so ; and the earth's curvature is less in those northern regions than between Ireland and Newfoundland. The distances were (varying a little according to landing-places selected) approximately :—

		miles.
From the North of Scotland to Faröe Islands		225
„ the Faröe Islands to Iceland		280
„ Iceland to Greenland, S.W. Harbour ...		700
„ Greenland to Labrador		550
	Total ...	1,755

From the Electrician's point of view, these sub-divisions were extremély favourable, as compared with the long continuous length entailed by an Atlantic Cable between Ireland and Newfoundland. Then again, the soundings (except for a section between Greenland and Labrador) did not yield anything approaching the more southern depths.

But against these palpable advantages there was the engineering objection, which at first seemed insurmountable, that the Greenland coast was bound up by ice for a great part of the year, in addition to the risk of injury to the cable from the grounding of icebergs. This latter was of less moment, for it could be provided against by keeping the cable when approaching shore, in the middle of any inlet, and thus away from the shallow sides, where the icebergs "ground."

THE ATLANTIC CABLE

There was also the probable difficulty of obtaining a trained staff to work a line when laid to such inhospitable regions. However, having regard to the anxiety exhibited by many to get to the North Pole, and to remain for years in the coldest Arctic regions, this did not present an insuperable obstacle. The Faröe Islands and Iceland were fairly populated, and there were permanent Danish residents at Julianshaab, in Greenland ; while at the end of Hamilton Inlet, Labrador, the Hudson's Bay Company had one of their permanent posts. Of course, there was intense cold in winter, but the staff would not be condemned to such trials of fortitude and appetite associated with a diet of whale's-blubber, seal, and white bear—such as Nansen, and other Arctic explorers before him, have since undergone. Their comforts, too, would be about as good as in the lone station on that bay in Newfoundland, named — not inaptly — " Heart's Content," for as Cowper puts it :—

> Where to find that happiest home below ?
> Who can tell—when all pretend to know !
> The fur-clad fisher of yon frozen zone
> Boldly proclaims that happiest spot his own ;
> Extols the treasures of his stormy seas,
> And his long nights of revelry and ease.

This bold project—with a route across the coldest and iciest regions of the Atlantic—had been originally brought to the notice of the Danish Government by Mr. Wyld, the geographer, even before the Atlantic Telegraph Company had been established. It was again introduced in a different form by Colonel T. P. Shaffner, an American electrician of some note. Colonel Shaffner, who had been the pioneer of telegraphs in the Western States, published his opinion early in 1855 against so long a circuit as the direct Atlantic line in the following words :—

"I do not say that a galvanic, or magnetic, electrical current can never be sent from Newfoundland to Ireland ; but I do say that, with the present discoveries of science, I do not believe it practicable for telegraphic service."

In this, of course, he proved to be mistaken ; nevertheless he made a strong case of the series of short stages geographically afforded by the North Atlantic deviation. After the 1858 cable had ceased working, to back up his belief in the advantages of the route—which he characterised as having "natural stepping-stones which Providence had placed across the ocean in the North"—he actually chartered a small sailing vessel, and—*with his family on board*—put forth from Boston on the 29th August, 1859, for the purpose of making the preliminary survey. He landed at Glasgow in November of that year; and presented to the public the results of his voyage, at a meeting over which Dean Buchanan presided.

THE ATLANTIC CABLE

During the voyage Colonel Shaffner sounded the deep seas to be traversed between Labrador and Greenland, and between Greenland and Iceland. He found a firm supporter in Mr. J. Rodney Croskey, of London, who advanced the "caution" money to the Danish Government for the concessions requisite in the Faröes, Iceland, and Greenland.[1]

Colonel Shaffner had been on terms of friendship for years with Sir Charles Bright and his brother, who had both contributed considerably to his *Telegraph Manual* published in the United States. Thus, after this preliminary work, he and Mr. Croskey discussed the matter with Sir Charles who—though rather sceptical at the outset—soon saw its feasibility, and entered heartily into the project as its technical adviser.

The first point was to convince the public that there were no insuperable difficulties in the way, by further surveys and soundings of a detailed character, so as to ascertain the inequalities of the bottom as well as the materials of which it is composed. These soundings should be undertaken officially, if possible.

In the course of the spring of 1860, Colonel Shaffner read a paper on the proposed North Atlantic Telegraph to the members of the Royal

[1] Mr. Croskey also subsequently found the bulk of the capital for the exploring expeditions.

Geographical Society, and every assistance was rendered, under the able direction of the President, Earl de Grey, Sir Roderick Murchison, and the Secretary, Dr. Norton Shaw. On the 15th of May, Lord Palmerston granted an audience to an influential deputation, headed by the Right Hon. Milner Gibson, M.P., and four other members of the House of Commons, to solicit the assistance of Government in sending out ships and officers to make the necessary official survey for ascertaining the practicability of the proposed route. The Premier appeared fully to appreciate the advantages of the north-about scheme, and in a very short time the Admiralty were directed to send out an expedition for the purpose of making the required survey.

The Admiralty selected for this duty Captain McClintock, R.N.,[1] an officer of great experience in the navigation of the Arctic seas, and H.M.S. *Bulldog* was placed under his command. This distinguished officer was directed to take the deep-sea soundings, and he sailed from Portsmouth on his mission in June, 1860. In the meantime, the promoters of the enterprise purchased the *Fox*—the steam yacht formerly employed in the successful search for the remains of the Franklin expedition—and fitted her out for the purpose of making surveys of the landing-places of the respective cables.

[1] Later Admiral Sir Leopold McClintock, K.C.B., LL.D., F.R.S.

The *Fox* was placed under the command of Captain Young,[1] of the mercantile marine, an officer well known for his distinguished labours under McClintock in the Franklin search. At the same time, Dr. John Rae, F.R.G.S., an intrepid Arctic explorer, volunteered his services to join the *Fox*, and take charge of the overland expeditions in the Faröe Isles, Iceland, and Greenland. Colonel Shaffner, as concessioner—besides two delegates on the part of the Danish Government, Lieutenant Von Zeilau and Arnljot Olafsson — also accompanied the *Fox* expedition, to take part in the necessary surveys.

Before the departure of the *Fox*, which sailed on July 18th, 1860, Her Majesty the Queen, the Prince Consort, and other members of the Royal Family, honoured the enterprise by a visit to that vessel, while lying off Osborne, and took a lively interest in the details of the expedition.

After the Royal Visit, Sir Charles Bright, with other promoters and friends, saw the party off with many hearty good wishes.

On the return of the expedition, Sir Leopold McClintock wrote an able letter to Sir Charles thoroughly favouring the route as perfectly practicable, pointing out that the ice would not really prove a difficulty, and strongly approving of the orignal intention of a land line across Iceland to Faxe Bay, "as by so doing you will avoid the

[1] Now Sir Allen Young, C.B.

only part of the sea where submarine volcanic disturbances may be suspected."[1]

The project naturally attracted a great deal of attention in the Press. The *Mechanic's Magazine* —a technical journal standing by itself at that time and always of high repute—contained a leading article on December 21st, 1860, which, on account of its general interest in this connection, is given in full in the Appendices to this volume.

The results of the voyages of H.M.S. *Bulldog* and the steam-yacht *Fox* were brought before a crowded meeting of the Royal Geographical Society on January 28th, 1861, when Sir Leopold McClintock gave the first public account of his exceedingly numerous and careful surroundings along, and in the vicinity of, the proposed course of the cable, interspersed with many useful remarks and hints as to ice, the best time for laying the line, etc., as well as the probable sphere of volcanic action in and off the South of Iceland.

Then followed an exhaustive paper by Sir Charles Bright, giving a synopsis of Captain Young's report on his voyage in the *Fox*, including the examination of various estuaries and harbours, so as to enable a decision to be arrived at as to the best landing places, the climatic conditions, etc.

[1] Letter from Captain Sir F. L. McClintock, R.N., to Sir Charles Bright, 6th, December, 1860. See Appendices at end of this volume.

THE ATLANTIC CABLE

From both sets of soundings it was shown that, as a rule, the bottom was of ooze or shelly. Dr. Wallich, the naturalist of the expedition, had brought up brightly coloured star-fish from depths of over a mile; whereas it had previously been believed that nothing could possibly live under such an enormous pressure of water.

In concluding his paper, Sir Charles made the following remarks :—

Having thus presented to the Society some of the most valuable and interesting portions of Captain Young's report, I have only to observe that the result of the recent survey has been to remove from my mind the apprehensions— which I previously entertained in common with many others—as to the extent and character of the difficulties to be overcome in carrying a line of telegraph to America by the northern route.

Prior to the despatch of the surveying expedition we had no knowledge of the depth of the seas to be crossed, with the exception of the few soundings obtained by Colonel Shaffner in 1859 ; and our information as to the nature of the shores of Greenland in regard to the require- ments of a telegraphic cable was equally small.

These points are of vital importance to the prospects of the North Atlantic route, and the survey has placed us in possession of satisfactory particulars respecting them. The soundings taken by Sir Leopold McClintock will be a guide in the selection of the most suitable form for the deep sea lengths of the cable, while the information fur- nished by Captain Young will direct the construction of the more massive cables to be laid in the inlets of the coast.

It is not necessary to determine upon the precise landing-places, and other points of detail in connection with the enterprise, at the present time. But the promoters of the undertaking have received ample encouragement from the survey—and from the testimony of competent and experienced voyagers and sojourners in the countries to which the line is to be carried—to warrant them in proceeding with their labours with renewed vigour and confidence. When they have achieved that success which their perseverance and energy deserve, I am sure they will always gratefully remember that their endeavours at the stage of their operations, which is now under discussion, would have been very much less productive of good results but for the patriotic foresight of Lord Palmerston in ordering the *Bulldog* on her late successful service. We must also be most thankful for the assistance of Sir Leopold McClintock, Captain Young, Dr. Rae, and the Commissioner appointed to accompany the *Fox* by the Danish Government—as well as others who took part in the cause—whose patience and devotion to their self-imposed work has been above all praise. Nor can those interested in this important undertaking forget the great help rendered to them by the Royal Geographical Society."[1]

Then came a highly instructive paper by Dr. Rae. He gave a number of interesting particulars of his land surveys, the population, price of food, wages, etc. He also described the ride of the *Fox* party across Iceland, whilst mak-

---

[1] Bright's paper as above will be found given at full length in the Appendices at end of this volume.

ing important suggestions as to the route for the land line with a view to avoiding the geysers.

These papers were followed at the next meeting of the Geographical Society by an exhaustive discussion, at which Lord Ashburton, Admiral Sir Edward Belcher, Captain (afterwards Rear-Admiral) Sherard Osborn, R.N., C.B., Mr. John

THE NORTH ATLANTIC EXPLORING EXPEDITION, 1860.

Ball, F.R.S., and various gentlemen of Arctic expedition fame, spoke favourably of the project.[1]

Numerous photographs had been taken during the expedition by Mr. J. E. Tennison Woods;

[1] It was here that Sir William Fothergill Cooke took occasion to express the pride he felt in Sir Charles having been a pupil of his; and he expressed himself similarly at various times in public.

and—from sketches by Lieut. J. E. Davis, R.N. —Captain R. B. Beechey, R.N., made a beautiful oil painting of the party, including some of the Esquimaux, on the occasion of landing to explore the inland ice at Igaliko Fiord. This picture was in the possession of Sir Charles up to the time of his death in 1888.[1]

At this time however (1861), there was still too much discouragement owing to the stoppage in working of the first Atlantic cable, and the still more disastrous failure of the Red Sea and Indian lines, *which had been laid in sections*, besides the loss of other cables in the Mediterranean. Moreover, there were those who still feared the ice-floes; and in the end, the public did not respond sufficiently. Thus, after all, what came to be styled the "Grand North Atlantic Telegraph" project, which had been worked out with so much trouble and expense, was never actually realised.

Another scheme which attracted some attention about the same time was described as the "South Atlantic Telegraph." This was for a very long length of cable between the south of Spain and the coast of Brazil, touching at Madeira, the Canary Islands, Cape de Verde Isles, Don Pedro and Fernando de Noronha Isles on the way, and stretching out to the West Indies and the United States.

---

[1] The reproduction given here is from a photograph kindly lent by Sir Allen Young.

Then there was a project—concerning which Sir Charles was also consulted—for a cable on an intermediate route from Portugal to the Azores, and thence to America, *viâ* Bermuda and the Southern States.

Being, however, to a great extent foreign in their scope, these latter schemes found little favour with those in this country, who were by way of promoting such enterprises.

Section XV

The 1865 *and* 1866 *Cables*[1]

Though their cables had ceased to work, the Atlantic Telegraph Company was kept afloat by the promoters, whilst Mr. Lampson as vice-chairman, and Mr. Saward as secretary, were doing all that could be done to keep its object constantly before the public, in the hopes of raising fresh funds.

In 1862 the Government were prevailed on to despatch H.M.S. *Porcupine* to further examine the

---

[1] It should be observed that a considerable interval of time occurred between the events just dealt with and those forming the subject of the present section of this chapter. It was thought best to depart from order of date here and tell the story of early Atlantic Telegraphy (in accordance with the sub-title of this book) in a consecutive manner. The intervening period is accounted for, so far as our object is concerned, in the earlier part of Vol. II.

ocean floor 300 miles out from the coasts of Ireland and Newfoundland respectively.

It took a considerable time to get together the full amount of capital required for another Atlantic cable, for this could only be done gradually. The great civil war in America stimulated capitalists to renew the undertaking. One of the main advantages adduced was—on this occasion, as before—the avoidance of misunderstandings between the two countries. Another—intended by Mr. Cyrus Field as a special inducement to his fellow countrymen—was the improvement of the agricultural position of the United States, by extending to it the facilities, already enjoyed by France, of commanding the foreign grain markets. On this account, the project was warmly supported by the Right Honourable John Bright, M.P., and other eminent Free Traders.

Mr. Field, however, met with as little success in obtaining pecuniary support in the States, as he had in connection with the previous line. His brother, Mr. H. M. Field, writes :—

The summer of this year (1862) Mr. Field spent in America, where he applied himself vigorously to raising capital for the new enterprise. To this end he visited Boston, Providence, Philadelphia, Albany, and Buffalo, to address meetings of merchants and others. He used to amuse us with the account of his visit to the first city, where he was honoured with the attendance of a large array of " the solid men of Boston," who listened with an attention that was most flattering to the pride of the speaker addressing such an assemblage in the capital of his native State.

There was no mistaking the interest they felt in the subject. They went still further: they passed a series of resolutions, in which they applauded the projected telegraph across the ocean as one of the grandest enterprises ever undertaken by man, which they proudly commended to the confidence and support of the American public. After this they went home feeling that they had done the generous thing in bestowing upon it such a mark of their approbation. *But not a man subscribed a dollar!*

In point of fact, as before, the cable of 1865— as well as that of 1866—was provided for out of English pockets.[1]

Let us now substantiate this statement by a cursory glance at events. Mr. Thomas Brassey, M.P., was the first to be appealed to in this country, and he supported the venture nobly. Then Mr. Pender[2] was applied to, and here also substantial aid was forthcoming. Both these gentlemen had joined the Board of the Telegraph Construction and Maintenance Company which had just been formed (in April, 1864), as the result of a definite amalgamation of the Gutta-percha Company, and Messrs. Glass, Elliot & Company. Mr. Pender, who had been

[1] Both the second and third Atlantic cables may, in fact, be described as "contractor's affairs"; inasmuch as the responsibility rested with the contractors and their staff from start to finish.

[2] Afterwards Sir John Pender, G.C.M.G., M.P.

largely instrumental in effecting this combination, became the first chairman.

Shortly after the first Atlantic cable was laid, Messrs. Glass, Elliot & Co. availed themselves of the services of Mr. Canning and Mr. Clifford, whose engagements on Charles Bright's staff for the " Atlantic " Company had ceased. Thus, with an additional staff of electricians, they had placed themselves in a position to undertake direct contracts for laying, as well as manufacturing, submarine telegraphs. They had, indeed, carried out work of this character in the Mediterranean during the year 1860; and on the amalgamation of the two businesses above mentioned into a limited liability company, their position was still further strengthened.

The capital raised for the new cable by the Atlantic Telegraph Company (with fresh blood added to it) was £600,000; and by agreeing to take a considerable proportion of their payment in " Atlantic " shares, the contractors, now the Telegraph Construction Company,[1] practically found more than half of this amount.

It will be seen that the new cable was to be an expensive one as compared with with that of 1857–

---

[1] This firm had previously (as Glass, Elliot & Co.) been selected to undertake the entire work on account of their experience, as well as owing to the considerable interest and pains they had evinced in the undertaking throughout.

58. It was the outcome of six years' further experience, during which several important lines—dealt with in Vol. II.—had been laid. It also followed upon the exhaustive Government enquiry to which we have already alluded.

The actual type, adopted on the recommendation of Sir Charles Bright and other engineers who were consulted, was much the same in respect to the

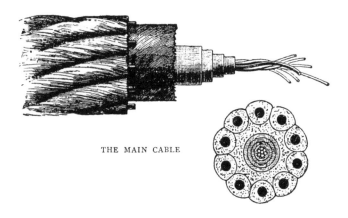

THE MAIN CABLE

conductor and insulator [1] as that which Sir Charles had suggested for the previous Atlantic line, on which occasion, it will be remembered, his recommendation was not adopted. If it had been, it would have been a far larger core than anything dreamt of at that time. The armour provided for

[1] 300 lbs. copper to 400 lbs. gutta-percha per nautical mile.

Bright was also specially consulted regarding the general estimate.

the present insulated and yarn-served heart, or core, was precisely similar to Sir Charles' Government specification of May, 1859, for the proposed cable from Falmouth to Gibraltar. It consisted of a combination of iron and hemp,[1] each wire being enveloped in Manilla yarns.[2] The object of encasing the separate wires in hemp was (1) to protect them from rust due to exposure to air and water, and (2) to reduce the specific gravity of the cable, with a view to rendering it more capable of supporting its own weight in water. This form of cable—bearing a stress of about eight tons[3]—was considered by most of the authorities at that period to perfectly fulfil the conditions required for deep-sea lines.[4]

[1] Ever since the difficulties experienced in maintaining control during the laying of Mr. Brett's lines in the Mediterranean, the claims of light cables—without any iron—had been urged by various inventors and engineers, and this was a sort of a compromise.

[2] Mr. Edwin Payton Wright, of the firm of John and Edwin Wright, had a similar—and very ingenious—device for ordinary ropes of increased strength, and this patent was ultimately worked to.

[3] The increased breaking strain here afforded over that of the first Atlantic line was partly due to the great improvement made in the manufacture of iron wire during the interval.

[4] Experience has since taught us, however, that such a type lacks durability, owing to the rapid decay of the hemp between the iron wires and the sea. When the

THE ATLANTIC CABLE

It was determined that this time the cable must be laid in one length (with the exception of the shore ends), by a single vessel. There was but one ship that could carry such a cargo. This ship was the *Great Eastern*—the conception of that distinguished engineer, Isambard Kingdom Brunel. She was in course of construction, by the late Mr. Scott Russell, at the time of the first cable, and Charles Bright had cordially joined with Brunel in his regrets that she was not then available. An enormous craft of 22,500 tons, she did not prove suitable at that time as a cargo boat; and the laying of the second Atlantic cable was the first piece of useful work she did, after lying more or less idle for nearly ten years.[1] It is sad to think of the way this poor old ship was metaphorically passed from hand to hand. Even at this period, three separate companies had already been formed one after another to work her. As promoter and chairman of one of these, Mr.

hemp has once decayed a bundle of loose wires are left, which by exposure all round soon become seriously reduced and weakened. Moreover, this pattern was found afterwards to be unsuitable on account of a broken wire being liable to stab the insulation—an accident which could scarcely happen to a close-sheathed type.

[1] The *Great Eastern* in point of size was only a little before her time. In the present day, with improved engines, she could be usefully and profitably employed, had she not been broken up.

Gooch, C.E. (afterwards Sir Daniel Gooch, Bart., M.P.), took an active part in arranging that she should be chartered for this undertaking. Hence it was that he became a prominent party in the enterprise, with a seat on the board of the Telegraph Construction Company.

All the cable machinery was fitted to the *Great Eastern*, on behalf of the Telegraph Construction Company, by Mr. Henry Clifford (to the designs of Mr. Canning and himself), assisted by Mr. S. Griffith. In the main principles, the apparatus employed was similar to that previously adopted on the *Agamemnon* and *Niagara*.[1] There were, however, several modifications introduced, as the result of the extra experience gained during the seven years interval. The main point of difference was the further application of jockeys, in a more complete form, to the *Great Eastern* paying-out gear. All the machinery for the present undertaking was constructed and set up by the famous firm of engineers, Messrs. John Penn & Son, of Greenwich.

[1] This general similarity is referred to in the complete account of the 1865 and 1866 machinery, given by Mr. Elliot (afterwards Sir George Elliot, Bart., M.P.) in the course of a paper read before the Institution of Mechanical Engineers in 1867. This gentleman—as partner in the firm of Glass, Elliot & Co.—became a director of the Telegraph Construction Company on its formation, ultimately occupying the position of chairman up to the time of his death in 1893.

THE ATLANTIC CABLE

At length, all the cable having been manufactured, and shipped from the Greenwich Works, the *Great Eastern*, under the command of Captain (afterwards Sir James) Anderson,[1] left the Thames on July 23rd, 1865, and proceeded to Foilhommerum Bay, Valentia. Here she joined up her cable to the shore end [2] which had been laid a day earlier by s.s. *Caroline*, a small vessel chartered and fitted up for the purpose. The great ship then started paying out as she steamed away on her journey to America, escorted by two British men-of-war, the *Terrible* (Captain Napier) and the *Sphinx* (Captain Hamilton).

On behalf of the contractors—the Telegraph Construction and Maintenance Company — Mr. (now Sir Samuel) Canning was the engineer in charge of the expedition, with Mr. Henry Clifford as his chief assistant. As we have seen, both these gentleman had been engaged with Sir Charles Bright on the first Atlantic expedition, and had

---

[1] Captain Anderson had the reputation of possessing great skill in the handling of a ship. He was at the time in the service of the Cunard Steamship Company, by whose permission he joined the expedition.

[2] This—somewhere near thirty miles in length—had been made by Mr. W. T. Henley, of North Woolwich. It had an additional outer sheathing of iron strands, each strand being composed of three stout wires, bringing the weight up to as much as twenty tons per mile.

had much experience, alike in cable work and mechanical engineering. There was also on the engineering staff of the contractors, Mr. John Temple (formerly Bright's secretary and assistant engineer) as well as Mr. Robert London. Mr. C. V. De Sauty acted as chief electrician, assisted by Mr. H. A. C. Saunders, and several others. By arrangement with the Admiralty, Staff-Commander H. A. Moriarty, R. N., acted as the navigator of the expedition. Captain Moriarty was possessed of great skill in this direction—a fact which had been made clear in the previous undertakings.

Sir Charles Bright did not accompany this expedition. As will be seen in a subsequent chapter (of Vol. II.), he was at the time deeply engaged in political matters. Indeed, his visits to Greenwich had been of late largely associated with the General Election; and these visits terminated in his being returned for that borough.

The Atlantic Telegraph Company was represented on board by Professor Thomson and Mr. C. F. Varley, as electricians, the former acting mainly as scientific expert in a consultative sense. Both Mr. Field and Mr. Gooch accompanied the expedition, the former as the initial promoter of the enterprise, and the latter on behalf of the *Great Eastern* Company. Representing the Press, there were also on board Dr. W. H. Russell, the well-known correspondent of *The Times*, as the historian of the enterprise;[1] and Mr. Robert Dudley, an artist of repute, who produced several excellent pictures of the work in its different stages, as well as articles for the *Illustrated London News*.[2]

Inasmuch as Bright was not on board we will not attempt to give a detailed account of the trip. It suffices to say that several mishaps occurred during the laying. A number of unsuccessful attempts were made to recover the cable after it had broken in deep water when endeavouring to haul back a fault. Ultimately the ships had to return home, on August 11th, without having completed their work.

[1] This famous war correspondent, author, and general newspaper writer, has since become Sir William Howard Russell, LL.D.

[2] Detailed and stirring accounts of the events of this expedition also appeared subsequently in *Blackwood's Magazine*, in *Cornhill*, and in *Macmillan's*. The former was written by Mr. Henry O'Neil, A.R.A., and the latter by Mr. John C. Deane, both of whom were eye-witnesses aboard the *Great Eastern*. Mr. O'Neill also brought out an illustrated comic journal during this and the following expedition, issued at periodic intervals, which was a source of much amusement to those who had leisure for perusing it. Still more mirth-provoking was an "extravaganza," written by Nicholas Woods and J. C. Parkinson, on the subject, performed on board on the completion of all the work in 1866. Both these contributions were afterwards published in booklet form, and are much treasured by the parties caricatured therein.

SIR CHARLES TILSTON BRIGHT

Second and Successful Attempt, 1866.—The results of the last expedition, disastrous as they were from a financial point of view, in no wise abated the courage of the promoters of the enterprise. During the heaviest weather the *Great Eastern* had shown exceptional "stiffness"; whilst her great size and her manœuvring power (afforded by the screw and paddles combined) seemed to show her to be the very type of vessel for the kind of work in hand. The picking-up gear, it was true, had proved insufficient;[1] but with the paying-out machinery no serious fault was to be found. The feasibility of grappling in mid-Atlantic had been demonstrated, and they had gone far towards proving the possibility of recovering the cable from similar depths.[2]

[1] This was specially pointed to by the representatives of the "Atlantic" Company when considering the carrying out of the agreement by the contractors.

[2] It may be mentioned in passing also that Professor W. Thomson had in the interval delivered an Address before the Royal Society of Edinburgh, on "The Forces concerned in the Laying and Lifting of Deep Sea Cables" (see *Proc. Roy. Soc.*, December, 1865).

He had previously contributed an article to the *Engineer* relative to the catenary formed by a submarine cable between the ship and the bottom, during submergence, under the influence of gravity, fluid friction, and pressure. In this communication Professor Thomson pointed out that the curve becomes a straight line in the case of no tension at the bottom—the normal condition, in fact, when paying-out.

THE ATLANTIC CABLE

To overcome financial difficulties, the Atlantic Telegraph Company was, practically speaking, amalgamated with a new concern, the Anglo-American Telegraph Company, which was formed, mainly by those interested in the older business, with the object of raising fresh capital for the new and double ventures of 1866. The ultimate capital of this Company amounted (as before) to £600,000.

In raising this, Mr. Field first secured the support of Mr. (afterwards Sir Daniel) Gooch, M.P., then chairman, and previously locomotive superintendent, of the Great Western Railway Company, who, after what he had seen on the previous expedition, promised, if necessary, to subscribe as much as £20,000. On the same conditions, Mr. Brassey expressed his willingness to bear one-tenth of the total cost of the undertaking. Ultimately, the Telegraph Construction Company led off with £100,000, this amount being followed by the signatures of ten directors interested in the contract (as guarantors) at £10,000 apiece. Then there were four subscriptions of £5,000, and some of £2,500 to £1,000, principally from firms participating in one shape or another in the sub-contracts. These sums were all subscribed before even the prospectus was issued, or the books opened to the public. The remaining capital then quickly followed.

The Telegraph Construction Company, in under-

SIR CHARLES TILSTON BRIGHT

taking the entire work, were to receive £500,000 for the new cable in any case; and if it succeeded, an extra £100,000. If both cables came into successful operation, the total amount payable to them was to be £737,140.

It was now proposed not only to lay a new cable between Ireland and Newfoundland, but also to repair and complete the one lying at the bottom of the sea. A length of 1,600 miles of cable was ordered from the Telegraph Construction and Maintenance Company. Thus, with the unexpended cable from the last expedition, the total length available when the expedition started would be 2,730 miles, of which 1,960 miles were allotted to the new cable, and 697 to complete the old one, leaving 113 miles as a reserve.

The new main cable was similar to that of the year before.

The shore-end cable determined on in this case was of a different description. It had only one sheathing, consisting of twelve contiguous iron wires of great individual surface and weight; and outside all a covering of tarred hemp and compound. The part of this cable which was intended for shallow depths was made in three different types. Starting from the coast of Ireland, eight miles of the heaviest was to be laid, then eight miles of the intermediate, and lastly fourteen miles of the lightest type, making thirty miles of shoal-water cable on the Irish side. Five miles

386

of shallow-water cable, of the different types named, were considered sufficient on the New-foundland coast.

The previous paying-out machinery on board the *Great Eastern* was altered to some extent by Messrs. Penn to the instructions of Messrs. Canning & Clifford.

THE SHORE-END CABLE

Though different in details, the main improvement over the 1865 gear consisted in the fact that a 70 horse-power steam-engine was fitted to drive the two large drums in such a way that the paying-out machinery could be used to pick up cable during the laying, if necessary,[1]

---

[1] As has been shown, a like provision was made for the expedition of 1858.

and thus avoid the risk incurred by changing the cable from the stem to the bows. This addition of Penn Trunk - engines, as well as the general strengthening of the entire machinery, was made in accordance with the designs of Mr. Henry Clifford.

The picking-up machinery forward, after the previous expedition, was considerably strengthened and improved with spur wheels and pinion gearing. This formed an exceedingly powerful machine, and reflected great credit on those who devised and constructed it.

Similar gear was fitted up on board the two vessels—s.s. *Medway* and s.s. *Albany*—chartered to assist the *Great Eastern.*

For the purpose of grappling the 1865 cable, twenty miles of rope were manufactured, which was constituted of forty-nine iron wires, separately covered with manilla hemp. Six wires so served were laid up strand-wise round a seventh, which formed the heart, or core, of the rope. This rope would stand a longitudinal stress of thirty tons before breaking.

In addition, five miles of buoy rope were provided, besides buoys of different shapes and sizes, the largest of which would support a weight of twenty tons. As on the previous expedition, several kinds of grapnels were put on board, some of the ordinary sort, and some with springs to prevent the cable surging and thus escaping whilst the grapnel was still dragging on the bottom : others, again, were fashioned like pincers, to hold (or jam) the cable when raised to a required height, or else to cut it only, and so take off a large proportion of the strain previous to picking up.

BUOYS, GRAPNELS, MUSHROOMS—AND MEN

Most of this apparatus was furnished by Messrs. Brown, Lenox & Co., the famous chain, cable, anchor, and buoy engineers, several of the grapnels being to their design, as well as the " connections."

The propelling machinery of the *Great Eastern*

had similarly received alteration and improvement in the intervals of the two expeditions. Moreover, the screw propeller was surrounded with an iron cage, to keep the cable and ropes from fouling it, as had been provided by Sir Charles Bright for the *Agamemnon* and *Niagara* in 1857.

The testing arrangements had been perfected by Mr. Willoughby Smith, in such a way that insulation readings could be continuously observed, even whilst measuring the copper resistance, or while exchanging signals with Valentia. Thus there was no longer any danger of a fault being paid overboard without instant detection. On this occasion, also, condensers were applied to the receiving end of the cable, having the effect of very materially increasing — indeed, sometimes almost doubling—the working speed.

On the 30th June, 1866, the *Great Eastern* —steaming from the Thames, followed by the *Medway* and *Albany*—arrived at Valentia, where H.M.SS. *Terrible* and *Racoon* were found, under orders to accompany the expedition. The *Medway* had on board forty-five miles of deep-sea cable in addition to the American shore-end.

The principal members of the staff acting on behalf of the contractors in this expedition were the same as in that of the previous year : Mr. Canning was again in charge, with Mr. Clifford and Mr. Temple as his chief assistants. In the electrical department, however, the Telegraph Construction Company had since secured the services of Mr. Willoughby Smith as their chief electrician, whilst he still acted in that capacity at the Wharf Road Gutta Percha Works. Mr. Smith, therefore, accompanied the expedition as chief electrician to the contractors. Captain James Anderson and Staff-Commander H. A. Moriarty, R.N., were once more to be seen on board the great ship, the former as her captain, and the latter as navigating officer. Professor Thomson was aboard as consulting electrical adviser to the Atlantic Telegraph Company,[1] whilst Mr. C. F. Varley was ashore at Valentia as their electrician. Sir Charles Bright was at this period serving on various committees of the House of Commons, as alluded to in Vol. II.; but Mr. Latimer Clark took up his quarters at Valentia to personally represent the firm of Bright and Clark as consulting engineers to the Anglo-American Telegraph Company, Mr. J. C. Laws and Mr. Richard Collett[2] being re-

---

[1] Though financially wrapped up with the new "Anglo" Company, the "Atlantic" continued in existence till as late as 1874.

[2] At a later period—after both the 1865 and 1866 cables were in working order—Mr. Collett actually sent a message from Newfoundland to Valentia with a battery composed of a copper percussion cap, and a small strip of zinc, which were excited by a drop of acidulated water—the bulk of a tear only! This was during some experiments carried out by Dr. Gould on behalf of the Astronomer-Royal (in

spectively aboard and ashore at the Newfoundland end representing the same firm. Mr. Glass, the managing director of the Telegraph Construction, was ashore at Valentia for the purpose of giving any instructions to his (the contractor's) staff on board, whilst Mr. Gooch and Mr. Field were on board the *Great Eastern* as onlookers and watchers of their individual interests. There was also on board Mr. J. C. Deane, the secretary of the Anglo-American Telegraph Company, whose diary proved of much use to *The Times* and other newspapers in the absence of Dr. Russell, who had so vividly and thrillingly described the events of the previous expedition.

On the 7th July the *William Cory*—commonly known as the *Dirty Billy*—landed the shore-end in Foilhommerum Bay, and afterwards laid out twenty-seven miles of the intermediate cable. On the 13th, the *Great Eastern* took the end on board, and, having spliced on to her cable on board, started paying-out. The track followed was parallel to that followed the year before, but about twenty-seven miles further north. There were two instances of fouls in the tank, due to broken wires catching neighbouring turns and flakes, and thus drawing up a whole bundle of cable in an apparently inextricable mass of kinks

concert with the Magnetic Telegraph Company), between Greenwich and Newfoundland, *viâ* the Atlantic cables, for the verification of longitudes in the United States.

and twists quite close to the brake drum. In each case the ship was promptly got to a standstill, and all hands set to unravelling the tangle. With a certain amount of luck—coupled with much care—neither accidents ended fatally ; and, after straightening out the wire as far as possible, paying-out was resumed. Fourteen days after starting, the *Great*

THE *GREAT EASTERN* AT SEA

Eastern arrived off Heart's Content,[1] Trinity Bay, where the *Medway* joined on and landed the shore-end partly by boats, thus bringing to a successful conclusion this part of the expedition. The total length of cable laid was 1,852 nautical

[1] This is situated on the opposite side of Trinity Bay to Bull Arm, where the 1858 cable had been landed, and not so far up. It was supposed to be even more protected than Bull Arm, from which it is some eighteen miles distant.

miles, average depth 1,400 fathoms. After much rejoicing [1] during the coaling of the *Great Eastern*,[2] the Telegraph Fleet once more put to sea, on August 9th.

Recovery and Completion of 1865 Cable.—It now remained to find the end of the cable lost on the 2nd August, 1865, situated about 604 miles from Newfoundland, to pick it up, splice on to the cable remaining on board, and finish the work so unfortunately interrupted the year before.

On August 12th, the *Great Eastern*, accompanied by s.s. *Medway*, arrived on the scene of action, where they joined H.M.S. *Terrible* (Captain Commerell [3]) and s.s. *Albany*, these vessels having left Heart's Content Bay a week in

[1] These were at first somewhat dampened by the fact that the cable between Newfoundland and Cape Breton (Nova Scotia) still remained interrupted, and that consequently the entire telegraphic system was not even now complete. However, in the course of a few days this line was repaired, and New York and the rest of the United States and Canada were put into telegraphic communication with Europe.

[2] Six steamers, laden with coal, had set out from Cardiff some weeks in advance to feed the *Great Eastern* on her arrival on the other side of the Atlantic.

[3] Now Admiral of the Fleet, Sir J. E. Commerell, V.C., G.C.B.

advance to buoy the line of the 1865 cable and commence grappling.

The plan decided on was to drag for the cable near the end with all three ships at once. The cable, when raised to a certain height, was to be cut by the *Medway*, stationed to the westward of the *Great Eastern*, so as to enable the latter vessel to lift the Valentia end on board.[1]

After repeated failures and many mishaps— partly owing to bad weather—the cable was hooked on the 31st of August by the *Great Eastern* (when the grapnel had been lowered for the thirtieth time), and picking up commenced in a very calm sea. When the bight of cable was about 900 fathoms from the surface, the grappling rope was buoyed. The big ship then proceeded to grapple three miles west of the buoy, and the *Medway* another two miles, or so, west of her again. The cable was soon once more hooked by both ships, and when the *Medway* had raised her bight to within 300

[1] This being, of course, before the days of cutting and holding grapnels as we now have them. These render it possible for a single ship to effect repairs, even where it is out of the question to recover the cable in one bight. Mr. Claude Johnson, Mr. F. R. Lucas, Mr. W. F. King, Mr. H. Benest, and Professor Andrew Jamieson, F.R.S.E., have—amongst others—devised special grapnels of this character.

fathoms of the surface she was ordered to break it. The *Great Eastern* having stopped picking up when the bight was 800 fathoms from the surface, proceeded to resume the operation as soon as the intentional rupture of the cable had eased the strain, which, with a loose end of about two miles, at once fell from 10 or 11 tons to 5 tons. Slowly but surely, and amid breathless silence, the long-lost cable made its appearance at last—for the third time—above water, a little before one o'clock (early morn) of 2nd September. Two hours afterwards the precious end was on board, and signals were exchanged with Valentia.

The recovered end was spliced on to the cable on board, and the same morning the *Great Eastern* started paying out about 680 nautical miles of cable towards Newfoundland.

On the 8th September, when only 13 miles from the Bay of Heart's Content, just after receiving a summary of the news in *The Times* of that morning, the tests showed a fault in the cable. The mischief was soon found to be on board the ship; and the faulty portion having been cut out, paying-out again proceeded, finishing the same day at eleven o'clock in the forenoon. The *Medway* immediately set to work laying the shore-end, and that evening a second line of communication across the Atlantic was completed. The total length of this cable, com-

THE *GREAT EASTERN* COMPLETING THE SECOND ATLANTIC CABLE

menced in 1865, was 1,896 miles; average depth, 1,900 fathoms.

The main feature and accomplishment in connection with the second and third Atlantic cables, of 1865 and 1866, was the recovery of the former in deeper water than had ever been before effected, and in the open ocean; just as in the first (1858) line it was the demonstration of the fact that a cable could be successfully laid in such a depth and worked through electrically.

Professor Thomson's reflecting apparatus for testing and signalling had been considerably improved since the first cable.

In illustration of the degree of sensibility and perfection attained at this period in the appliances for working the line, the following experiment is of striking interest:—Mr. Latimer Clark —who went to Valentia to test the cable for the " Atlantic" Company—had the conductor of the two lines joined together at the Newfoundland end, thus forming an unbroken length of 3,700 miles in circuit. He then placed some pure sulphuric acid in a silver thimble,[1] with a fragment of zinc weigh-

[1] Mr. Clark borrowed the thimble—which was a very small one—from Miss Fitzgerald, the daughter of the Knight of Kerry, living at Valentia. This gentleman— as has already been shown—evinced great interest in,

ing a grain or two. By this primitive agency he succeeded in conveying signals twice through the breadth of the Atlantic Ocean in little more than a second of time after making contact. The deflections were not of a dubious character, but full and strong, the spot of light traversing freely over a space of 12 inches or more, from which it was manifest that an even smaller battery would suffice to produce somewhat similar effects. This speaks well for the electrical components assigned to the two lines, and for the arrangements adopted in working them. It also shows the benefit derived from seven years' extra experience in manufacture, backed up by the previously-mentioned exhaustive Government enquiry thereon.

Notwithstanding the dimensions of the core, these cables were worked slowly at first, and at a rate of about eight words per minute. This, however, soon improved as the staff became more accustomed to the apparatus, and steadily increased up to fifteen and even seventeen words per minute on each line, with the application of condensers.

Unfortunately both these cables broke down a few months later, and one of them again during the following year. These faults were localised [1]

and offered every assistance in furtherance of, the Atlantic cable enterprise from the very beginning.

[1] The above location was performed by a method

with great accuracy from Heart's Content by Mr. F. Lambert, on behalf of Messrs. Bright and Clark, the engineers to the " Anglo-American " Company.[1] However, unlike the 1858 line, these last cables had not been killed electrically; and being worthy of repairs, they were maintained for a considerable time.

The success attending the above important operations gave rise to a new era in submarine telegraphy. The period of first attempts was virtually over.

On the return of the 1866 Expedition a banquet was given to the cable layers by the Liverpool Chamber of Commerce as soon as the *Great Eastern* was once more safely moored in the Mersey.

The following extracts from The *Morning Star* will be of some interest here :—

" The decorations assumed an emblematic character and were peculiarly appropriate to the event which was being celebrated. From the centre of the room there depended the grapnel by which the previous line was recovered from

based on Charles Bright's patent of 1852, already referred to.

[1] In some instances the faults were found to be due to grounding icebergs at the entrance to Trinity Bay.

the bed of the ocean, a piece of the cable itself and the grapnel chain. Then around the room were two lines of the cable supported by gilded grapnels, a profusion of sea-weed being entangled about the lines. The principal mirrors were surmounted by trophies of flags: those over the mirror at the rear of the President consisting of English and American flags, and those over the principal side mirrors being flags of all nations.

"A line of telegraph was extended from the British and Irish Magnetic Telegraph Company's Office, at the Liverpool Exchange, to the banqueting room; and as a practical illustration of the working of the cable a message was despatched to Washington, besides communications by the telegraph being read from Newfoundland.

"The Chair was occupied by the Right Hon. Sir Stafford Northcote, Bart.,[1] President of the Board of Trade. The following were amongst the invited guests: the Right Hon. Lord Stanley, M.P., Secretary of State for Foreign Affairs; the Right Hon. Lord Carnarvon; the Right Rev. the Lord Bishop of Chester; the Right Hon. W. E. Gladstone, M.P.; Sir Charles Bright, M.P., original projector of the Atlantic Cable, and engineer to the Anglo-American Telegraph Company; Prof. W. Thomson, electrical adviser to the Atlantic Telegraph Company; Mr.

[1] Afterwards the first Earl of Iddesleigh, G. C. B.

THE ATLANTIC CABLE

Latimer Clark, co-engineer with Sir C. Bright; Mr. R. A. Glass, managing director to the Telegraph Construction Company, (contractors); Mr. Samuel Canning, engineer to the contractors; Mr. Henry Clifford, assistant engineer to the contractors; Mr. Willoughby Smith, electrician to the contractors; Captain James Anderson, commander of the *Great Eastern*; Mr. William Barber, chairman of the Great Ship Company; Mr. John Chatterton, manager of the Gutta Percha Works; Mr. E. B. Bright, Magnetic Telegraph Company; Mr. T. B. Horsfall, M.P.; and Mr. John Laird, M.P.

After proposing toasts to Her Majesty the Queen, to the President of the United States, and to the Prince of Wales, the Chairman (Sir S. Northcote) again rose amidst applause and said it was a maxim of a great Roman poet that a great work should be begun by plunging into the middle of the subject. He would therefore do so by proposing a toast to the projectors of the Atlantic Telegraph — Sir Charles Bright and Mr. Cyrus Field, Mr. J. W. Brett having since unfortunately died. When they came in after years to relate the history of this cable, they would find many who had contributed to it; but it would be as impossible to say who were the originators of the great invention, as it was to say who were the first inventors of steam. He begged to couple with the toast the name of Sir Charles Bright as, perhaps, the foremost representative from all points of view, up to the present time (applause). The greatest honour is due to the indomitable perseverance and energy of Sir C. Bright that the original cable was successfully laid, though—through no fault of his—it had but a short useful existence (great cheering).

Sir Charles Bright, M.P., after acknowledging the compliment paid to the "original projectors" and to himself personally, said that the idea of laying a cable across the Atlantic was the natural outcome of the success which was attained in carrying short lines under the English and Irish Channels, and was a common subject of discussion among those concerned in telegraph extension prior to the formation of the Atlantic Telegraph Company.

About ten years ago the science had sufficiently advanced to permit of the notion assuming a practical form. Soundings taken in the Atlantic between Ireland and Newfoundland proved that the bottom was soft, and that no serious currents or abrading agencies existed ; for the minute and fragile shells brought up by the sounding-line were perfect and uninjured.

There only remained the proof that electricity could be employed through so vast a length of conductor. Upon this point, and the best mode of working such a line, he had been experimenting for several years. He had carried on a series of investigations which resulted in establishing the fact that messages could be practically passed through an unbroken circuit of more than two thousand miles of insulated wire—a notion derided at that time by many distinguished authorities. Mr. Wildman Whitehouse—who subsequently became Electrician to the Company—had been likewise engaged. On comparing notes later, it was discovered that we had arrived at similar conclusions, though holding somewhat different views ; for his (Sir C. Bright's) calculations, using other instruments, led him to believe that a conductor nearly four times the size of that adopted would be desirable with a slightly thicker insulator. It was this type, which the new cables just laid had been furnished with.

In 1856, Mr. Cyrus Field—to whom the world was as much indebted for the establishment of the line as to any man—came over to England upon the completion of the telegraph between Nova Scotia and Newfoundland. He then joined with the late Mr. Brett and himself (Sir C. Bright), with the view of extending this system to Europe, and they mutually agreed—as also did Mr. Whitehouse later—to carry out the undertaking.

A meeting was first held in Liverpool, and in the course of a few days their friends had subscribed the necessary capital. So that in greeting those who had just returned from the last expedition—Mr. Canning, Mr. Clifford, Captain Anderson, and other guests of the evening— Liverpool was fitly welcoming those who had accomplished the crowning success of an enterprise to which at the outset she had so largely contributed (applause).

The circumstances connected with the first cable would be in the recollection of every one ; and although the loss was considerable, the experience gained was of no small moment. A few months after the old line had ceased to work, their chairman (Sir S. Northcote) consulted him on behalf of the Government as to the best form of cable for connecting us telegraphically with Gibraltar, and he (Sir C. Bright) did not hesitate to recommend the same type of conductor and insulator which he had himself before suggested for the Atlantic line—a higher speed being desirable. This class of conductor in the newly-laid Atlantic cable appeared likely to give every satisfaction, he was happy to say ; and the mechanical construction of the cable—also the same as that he had previously specified for the Gibraltar line—appeared to have admirably met some of the difficulties experienced in cable operations.

The credit attached to these second and third Atlantic cables must mainly rest with the Telegraph Construction Company (formerly Messrs. Glass, Elliot & Co.) and their staff, inasmuch as in this case the responsibility rested with them throughout. The directors—including Mr. Glass, Mr. Elliot, Mr. Gooch, Mr. Pender, Mr. Barclay, and Mr. Brassey—deserved the reward which they and the shareholders would no doubt reap.

To Mr. Glass—upon whom the principal responsibility of the manufacture devolved—the greatest praise was due for his indomitable perseverance in the enterprise. Then the art of insulating the conducting wire had been so wonderfully improved by Mr. Chatterton and Mr. Willoughby Smith that, nowadays, a very feeble electrical current was sufficient to work the longest circuits— an enormous advance on the state of affairs nine years previously.

Again, they must not forget how much of the success now attained was due to Prof. Thomson and his delicate signalling apparatus, the advantages of which have, since 1858, been more firmly established. Mr. Varley had also done most useful work since becoming electrician to the " Atlantic " Company. Moreover, he (Sir C. Bright) hoped the active personal services of his partner, Mr. Latimer Clark, would not be lost sight of.

It was satisfactory to find that the cables were already being worked to a very large profit. This would doubtless be quadrupled within a short period, when the land lines on the American side were improved (hear, hear, and applause).

With this commercial success—combined with the improvements introduced into submarine cables, and the

power of picking up and repairing them from vast depths
—there was a future for submarine telegraphy to which
scarcely any bounds could be imagined. A certain
amount had already been done; but China and Japan,
Australia and New Zealand, South America and the
West India Islands must all be placed within speaking
distance of England. When this has been accomplished—
but not till then—telegraphic engineers might take a short
rest from their labours and ask with some little pride—

Quæ regio in terris nostri non plena laboris?

(loud applause).

Then followed speeches from Lord Stanley,
the American Consul (on behalf of Mr. Cyrus
Field), and others.

Honours were subsequently bestowed on some
of the various gentlemen most immediately concerned in those, at last, wholly successful undertakings of 1865 and 1866.

As a natural sequence other Atlantic cables
followed—first of all in 1869 that hailing from
France—until now (in 1898) the North Atlantic
ocean alone is spanned by as many as twelve in
working order.

APPENDIX I

THE BRIGHT PEDIGREE

The material originally positioned here is too large for reproduction in this reissue. A PDF can be downloaded from the web address given on page iv of this book, by clicking on 'Resources Available'.

APPENDIX II

By the late Hon. Martin-Bladen-Edward Hawke, uncle to the Right Hon. Lord Hawke, and present Master of the Hunt. With Notes and Notices, by William Sheardown, Esq.

The time that has elapsed since the following ballad was written has rendered it difficult to describe the names of the persons, mentioned in the song of 1730, further than has been already done in the *Poems on Hunting*, by the Hon. Martin-Bladen Hawke.[1] The song is therefore now reprinted as there inserted. It may not, however, be out of place to give a short notice of Mr. Bright, of Badsworth, the founder of the Badsworth Hunt.

DONCASTER, *August* 5, 1862.   WM. SHEARDOWN.

MR. BRIGHT, OF BADSWORTH.

By an act of Parliament, 1652, Badsworth, which had been in the possession of Robert Dolman, was ordered to be sold, for treason against the Parliament and the people, and was purchased by Colonel Bright, of Carbrook, in the parish of Sheffield, an eminent officer in the Parliament army. His services are narrated in *Hunter's Hallamshire*, pp. 249, 250. He left the army in disgust, having been refused a fortnight's leave of absence, when at Newcastle, in 1650, marching with Cromwell's army into Scotland. He settled at Badsworth, and was twice High Sheriff for the county of York in 1654-55. He is supposed to have assisted in bringing about the Restoration, and was admitted as a baronet in 1660, having previously been knighted. He was four times married, and died 13th September, 1688. By his first wife he had a son, John, who died without issue; his daughter, Catharine, married Sir Henry Liddell, Bart., of Ravensworth Castle, in Durham. Thomas, the eldest son by this marriage, was the ancestor

---

[1] *Poems on Hunting*, by the late Hon. Martin-Bladen Hawke, containing "The Badsworth Hunt, 1730" (Bright's Hounds), "Sessay Wood, 1806" (Sir Mark Masterman Sykes's hounds), "Howell Wood, 1805" (The Earl of Darlington's hounds), etc., with a sketch of his sporting career, by Nimrod, published in 1842 ; out of print.

APPENDIX

of the first Lord Ravensworth ; John, the second son—and the originator and Master of the Badsworth Hunt—was made principal heir of his grandfather, on whose death he assumed the name and arms of Bright ; and was at one time M.P. for Pontefract; he died 12th October, 1735. Thomas, his son—the " Tom " of the song—and last male descendant, was buried at Badsworth, 7th May, 1739, leaving a daughter, Mary, who married 26th of February, 1752, the Marquis of Rockingham, who died without issue. By this marriage the manors of Badsworth, Billingley, and Ecclesal, with other lands, were added to the estates of the house of Wentworth. Badsworth Hall has since been inhabited by — Winn, Esq. (Nostel), the Earl of Darlington, Colonel Mellish, John Pate Nevile, Esq., Joseph Scott, Esq., and now is the property and residence of Richard Heywood Jones, Esq. Badsworth is once mentioned in the history of the civil wars :—

" On the 15th March, 1645, a party of the King's Horse from Pontefract Castle fell on Colonel Brandling's quarters at Badsworth, took sixty-seven prisoners (of whom thirteen were officers), 130 horses, and £1,000 in money."—*Mercurius Rusticus.*

" THE BADSWORTH HUNT: Descriptive of an excellent Fox Chase, as performed by the hounds of Mr. Bright, of Badsworth, in the year 1730."

> " Hark ! what loud shouts re-echo thro' the groves,
> He breaks away, shrill horns proclaim his flight,
> Each straggling hound strains o'er the lawn to gain
> The distant pack—'tis triumph all and joy ! "
>
> Somerville.

Ye huntsmen, give ear to my song,
    Who to Sussex steep hills do resort ;
I sing of a fox-chase so long,
    That you must allow it good sport.

It was in the time of the year
    When foxes could fly and were stout,
In Badsworth's gay hall did appear
    Of hunters a jovial rout.

Said the Master[1] o'ernight, " It is ten ;
    Call Slinger,[2] for I will to bed :
At five I will see you again ;
    Pray, Tom,[3] now remember your head."

---

[1] John Bright, Esq., of Badsworth Hall, near Pontefract, was the Master of the Hounds.
[2] Slinger was Mr. Bright's valet.
[3] Mr. Bright's son.

APPENDIX

At five, then, the Master arose;
　The rest, half-asleep, left their beds,
And hastily donn'd on their clothes;
　Tho' some of 'em felt heavy heads.

To cover they walk a foot's pace,
　Where the company all does appear,
But Harvey,[1] who lost all the chase,
　By taking twice leave of his dear.

It was just at the rise of the sun,
　To Barnsdale's great whin-bed they came—
So famous for many a run,
　So crowded for fox-hunters' game.

"Hoix, Truelove," said Jarvese, "my hound;"
　"Hey, Jumbler," Jack[2] quickly replied;
"Egad!" said Ben Sayle, "he is found;
　Hark! Duchess, who never yet lied."

"Hollo!" then away the pack goes:
　"Master Wilson, come on," says Tom Sayle.[3]
Kit[4] answers, "I'll gather these sloes,
　And then comb my nag's mane and tail."

Over Stapleton Lees to Wake Wood,
　Down to Balne, still up wind he doth fly;
But soon found, in spite of his blood,
　He must back again, else he must die.

From Grove Wood and sheer to Went Hill,
　Where a huntress came up to the cry:
Her voice was so sweet and so shrill,
　It must be Diana,[5] or Di.

From hence hied to Darlington Moor,
　Over Went and by Badsworth he goes.

---

[1] Mr. Harvey, grandfather to the present Lord Hawke, then lived at Womersley Park.

[2] The huntsman's name was A. Jarvese, and the whipper-in was called "Jack."

[3] Thomas and Benjamin Sayle were brothers, and resided at Wentbridge.

[4] Wilson then lived at Wakefield, and was generally styled "Kit Wilson of famous memory," as it is stated in the notes of the song published at the time.

[5] Diana Sayle, sister of Thomas and Benjamin Sayle.

APPENDIX

Oh ! Reynard, thy fate I deplore,
    For there lives the worst of thy foes.

Then up to the Hollins he ran,
    Where a ploughman he met in the face.
This lucky hit let in each man,
    Or few had been seen in at the chase.

The Master came up in his chair,
    Saw Danger hit off the default,
And said, " Had Ralph Elmsall [1] been there,
    Hey, Danger, he'd quite split his throat."

" Now, Rockwood," " Now, Delver," some cried ;
    " Now, Rival," " Now, Sempstress," again ;
Then Hall [2] his dog Rebel espied,
    And swore he led over the plain.

" Zounds ! " says Kitchingham, " Hall is foreswore,
    But *he'll* swear any man off his nag ;
See Tapster and six couples more ;
    He cannot blow wind in their bag."

'Squire Thomas came up to the head,
    And swore they were every one blind ;
" For see, my dog Juggler does lead,
    And Tippler is not far behind."

He then made for Hampole high wood,
    But found it too hot for his stay :
Smith [3] saw him, as watching he stood,
    And bid him make best of his way.

To Brodsworth he cunningly stole,
    And then sheer'd away to the Marr,
At the warren at Melton to hole :
    But Dawson [4] had put up a bar.

Over Don,[5] then, he hastens away ;
    On Conisbrough Cliff he relies.
O ! Renny, in vain is thy play ;
    For mountains put up, and thou dies.

---

[1] Ralph Elmsall lived at Thornhill, near Wakefield.
[2] Mr. Anthony Hall, of Wombwell.
[3] Smith lived at Brodsworth, and was the warrener.
[4] Dawson was the earth-stopper.
[5] " Over Don." This took place over the ferry at Sprotbro'.

APPENDIX

The boatman was luckily by;
The horsemen, with heart and good-will,
Got over, and presently spy
The hounds dancing over the hill.

Here Molly the lead she does take;
Oh! Roper, she doth so behave—
Tippler's blood thy dead corpse should awake,
And make thee jump out of thy grave!

For Edlington Wood then he flew:
Ere Edlington Wood he could reach,
They ran out of scent into view,
And Diamond laid hold of his breech.

"Whoo, hoop!" then Dick Sunderland cries;
Tom Atkinson[1] stood in amaze;
The company own'd with surprise
Such a chase they ne'er saw in their days.

Now, pray, my good people attend:
The chase it was thoroughly run;
Tom Bright was at Hampole Wood end
When the hounds they were crossing the Don.

"And why need you marvel at that,"
Says the Captain of Skelbrooke Hall;[2]
"Perhaps he is watching a cat—
A coney—or nothing at all."

Hence, Warmsworth, shall thy haughty spire
Our fame to posterity bear;
While Childers and Newby admire,
And Draper[3] with envy shall hear.

Now to Badsworth's roast beef let us hie,
Where we'll finish the day with delight;
We'll drink to fox-hunters and Di,
And fuddle our noses all night.

---

[1] Roper, Atkinson, and Sunderland were noted hunters in the county of Sussex.
[2] The name of the "Captain of Skelbrooke" was Brown.
[3] Childers, Newby, and Draper were all noted hunters.

APPENDIX III

THE following represents Charles Bright's Report, as engineer to the above Company, for the year 1853 : —

REPORT

I beg to report that the following important works have been completed since the last general meeting of the Company's shareholders. The underground trunk line of ten wires from London to Manchester and Liverpool was finished at the end of October, and has since continued in efficient operation.

The underground line of six wires from Liverpool to Carlisle, and thence to Portpatrick, was opened lately, and is working satisfactorily. The submarine cable of six wires between Great Britain and Ireland was successfully laid from Portpatrick, in Wigtonshire, to Donaghadee in May, and remains in the same perfect condition as I then reported to you.

The underground line from Donaghadee to Belfast has been finished a few days since and Belfast is now connected in direct circuit with Liverpool.

The underground line from Cork to Queenstown has been in operation since the commencement of the present year.

The line of pole telegraph from Belfast to Dublin, with branches to Armagh, Kells, Newtonards, and Holywood, along the county Down, Ulster, Belfast Junction, and Dublin and Drogheda railways, has been open for commercial business in connection with the cable since the beginning of January.

The same system from Dublin to Cork, with branches to Killarney and Carlow on the Great Southern and Western and Killarney Junction Railways, was completed and opened for public use at the same time.

The works connected with the construction of the foregoing lines have been well and faithfully executed by the several contractors and the Company's workmen, and completed with great despatch, though impeded to a considerable extent by circumstances affecting the supply of the chief materials employed.

The chief part of your lines are in the most satisfactory and perfect condition of efficiency, but some of the over-ground branches will be

APPENDIX

benefited by certain improvements as regards their insulation, which will be proceeded with immediately, when the season is sufficiently advanced.

Tenders have been accepted from responsible and experienced contractors for the erection of a line of telegraph upon the Waterford and Limerick Railway, and for six additional wires between Dublin and Belfast.

Agreements of a satisfactory nature have also been entered into with the Londonderry and Enniskillen and Coleraine Railways for the supply of four telegraphs to those lines. One hundred and three new instruments, and various machinery necessary to the proper working of the additional stations and increased business, have been provided during the year, and the apparatus of the Company is in good and efficient condition.

CHARLES T. BRIGHT.

APPENDIX III*a*

THE following was Charles Bright's Report for the year 1854 :—

ENGINEERS' REPORT

LIVERPOOL, *February* 15*th*, 1855.

GENTLEMEN,—

The following progress has been made with the works undertaken by you during the past year :—

Underground lines consisting of six wires have been laid between Dublin and Belfast, and from Liverpool to Preston, and a line of four underground-wires has been extended between Dumfries, Glasgow, and Greenock. These lengths constitute important links in the Company's system, and by their completion direct communication has been established between Liverpool and Glasgow, Belfast, or Dublin, in addition to the unbroken connection previously effected with the intermediate towns on these lines, and with London and Birmingham on the south.

An overground line of four wires has been erected between Londonderry and Enniskillen, and others are in course of formation that will connect Belfast with Ballymena, Coleraine, Londonderry, and Enniskillen. Dublin has been placed in communication with Limerick by means of the extension on the Waterford and Limerick Railway, and additional wires are being laid on the pole-line south of Dublin to meet the increased traffic arising from the new line.

Two short branches for railway purposes are also in progress on the Great Southern and Western and East Lancashire lines.

The above works, comprising over 3,000 miles of wire, have been carried out in the most improved and substantial manner, and within the estimates submitted to you prior to their being undertaken ; and when the lines now in hand are finished the system of the Company, as far as hitherto proposed, will be complete.

The temporary stations at Dublin, Belfast, and Greenock have been exchanged for more suitable and commodious offices, and new stations have been opened at Charing Cross and the Stock Exchange in London, Port Glasgow, Fleetwood, Southport, Londonderry, Enniskillen, Killarney, Limerick, and Newry.

APPENDIX

Offices are also about to be established at Mincing Lane and Chancery Lane in London, and at Waterford, Dundalk, Armagh, Ballymena, and Coleraine.

Improvements have been introduced in the instruments, by which the effective range of the Company's telegraphs has been considerably increased, and we have now in operation the longest circuits anywhere working constantly, and without variation from bad weather. In addition to the advantages we gain of greater celerity and less liability of error by reason of not having to repeat at intermediate points, the saving of staff otherwise employed at transmitting stations is an important gain to the Company.

The lines and apparatus have been in a good and efficient condition, and the submarine cable between Great Britain and Ireland continues in the same perfect state of insulation as reported last year, forming part of a circuit of 320 miles in continual use.

Arrangements have been made with responsible contractors for the maintenance of the underground wires between London and Manchester, and between Liverpool and Glasgow and Port Patrick, by which it is stipulated, and security provided, that the wires are to be kept up to a satisfactory standard of insulation, and returned to the Company in an equally efficient state at the expiration of ten years.

I am, gentlemen, your most obedient servant

CHARLES T. BRIGHT

APPENDIX IV

IN the early days our cousins across the water were a little behind us in telegraphy. Thus, in 1855, the authorities, through the agency of Colonel T. P. Shaffner (a well-known American telegraph engineer), addressed a number of questions to Charles Bright and his brother with a view to learning something about English systems.[1]

These questions, as propounded by Colonel Shaffner, are given here, and in subsequent pages the answers will be found. The whole matter is given *in extenso*, for it is believed that such a collection of practical points have never been put together before, and that they contain much that will be of interest in the present day—nearly fifty years later.

QUESTIONS.

1. Do the wires of your Company run over ground or under ground —and to what extent?

2. If on poles, what kind of timber do you find the most durable— and, if possible, please state about what age are the poles (or timber) thus employed?

3. Do you use any pitch, tar, or other matter on your poles, to increase their durability—and if so, what and how applied?

4. Please state what kind of insulators you use on your poles, and, if possible, please give a drawing of them or samples of each, with your opinion as to their fitness or faults—also their cost?

5. Please state the expense of your poles, and what is the cost for digging holes, raising the poles, the putting on of insulators, and placing the wires on the poles?

6. What kind of wire do you use, and where mostly manufactured —and what is the price for the same, per pound?

7. Do you use galvanized wire, and what are its advantages or disadvantages?

8. Do you solder the joints of your wire—if not, do you find any difficulties arising from oxidation at joints?

---

[1] To pursue the investigation, Colonel Shaffner visited England.

418

APPENDIX

9. Do you realise much difficulty in the use of either galvanized wire or other wire, on poles, from atmospheric electricity—on which the most, and at what seasons of the year?

10. Where you have wire on poles, do you find any difficulty arising from cross currents at the poles ; that is, the current passing from one wire to another at the poles?

11. Are your instruments ever affected by induced currents ; that is, the passage of the galvanic or magnetic electricity from one wire to another, by or through elements of nature other than material substances ; and if so, to what extent?

12. Do you ever suffer from what may be denominated "heat lightning"; and if so, to what extent?

13. Do you suffer from atmospheric electricity, either accompanied or not accompanied with thunder, and to what extent?   Also, how do you protect your instruments from harm?

14. Do you know of the burning of any property through the agency of the electric wires ; and if so, to what extent?

15. Are the telegraph poles ever struck and damaged by lightning ; and if so, how often in a year, averaging for a scale of 100 miles?

16. How high are your poles, and how many do you use per mile?

17. How many wires can you place upon one set of poles?

18. Do you find any difference in the working of the wires on the poles ; that is, the upper, middle, or lower wires?   If so, what is that difference in fair weather, warm or cold, wet or dry seasons, and in time of storm?

19. Do you fasten the wire at each pole ; and if so, how?

20. Do your wires often break ; and if so, what causes them to break?

21. How do you mend your breaks?   How many persons are required, and what is the mode you adopt to make the joint?

22. Do your operators usually go on the line to repair breaks, or other damages to the line?

23. Do you employ police to guard the line ; and what is the plan, or system, and the expense?

24. Does the snow in winter disturb the working of your lines on poles ; and if so, what are the remedies?

25. Does much ice form on your wires ; and do the wires break, caused by the weight of the ice?

26. Please state your mode of laying underground lines, and furnish drawings, or samples, if convenient?

27. Do you find any difference in the use of wires covered once or twice with gutta percha?

28. What are the difficulties presented in laying wires covered only with hemp, over the gutta percha? and please state the different modes, with their respective costs.

419

APPENDIX

29. What are the causes of breaks, and their frequency, with underground lines?

30. How do you discover the place of break in a subterranean line? Please give a drawing of the plan, and state the time usually required to make the repair. Please specify fully on the subject.

31. What is the cost of laying one, two, or more wires? giving the cost of labour, depth of ditch, and the plan in detail.

32. In case of much rock on the surface, do you blast; and how do you lay the wires?

33. Where there are marshes, how do you lay the wires?

34. Do you suffer from the upheaving of the earth, in case of frost in winter, and to what extent; and what are the remedies to avoid it?

35. What seasons of the year do your wires suffer the least?

36. What are your plans for crossing small streams?

37. Do you suffer from cross or induced currents from one wire to another in underground lines?

38. Did you find any difference whatever in the working of the wires, by their increased number in any combination from one to ten or more underground? If so, what is that difference?

39. What is the relative quantity of battery you use on underground lines compared with lines on poles?

40. What battery do you believe the best, and what quantity required for a distance of 100 miles? Please give the cost of the materials in items. Can you work more than one independent wire forming an independent circuit from the same battery? If so, how many—and by what arrangement and principle?

41. How often do you repair the battery; what is that repair, and its expense?

42. What are your plans for protecting your line and instruments from lightning?

43. What do you consider return currents; and to what extent do you find the existence of the same on both overground and underground lines? Please state all the points fully.

44. Have you discovered any difference in the time required in the transmission of a current on the overground or underground lines, or in submarine lines; and what are the facts respectively?

45. Have you found any advantages in the use of any given size wire for electric conductors, either over or underground; and what are they?

46. What is the difference in the practical use of a line of iron wire and one of copper, as far as you are able to judge?

47. Do your underground wires ever suffer from lightning?

48. Do you allow the Government any advantages in sending messages; and what are those advantages?

APPENDIX

49. Has the Government given any grants, appropriations, or other advantages or benefits to the telegraphs, either in law or in its use?

50. Please give the mode of receiving messages from the public, and the various checks placed upon the message, and the time thus employed, commencing at the reception at the counter, and ending with its delivery at the destination.

51. Do you ever send the messages not signed ; or when written on any other paper than your printed forms?

52. Why do you require persons to use your printed forms, and has that been the practice from the commencement of the telegraph? Please give two blanks, thus used, one filled in to illustrate the plan you follow, with all the explanations needed to enable a stranger to understand the same?

53. Please give a form of your register books, upon which you enter the messages you send and receive, with explanations of their use. Also copies of your rules and regulations as to the Company, and of working the line.

54. Please state the average salaries you pay for the respective officers required in your city and country offices.

55. Do you clothe your messengers?

56. Do you place any of your officers under oath or bonds, and is there advantages to the public or Company by so doing, and do the Government laws require it?

57. Are you often called upon to give copies of messages to persons, and do you retain copies in your office?

58. What is the cost of your printed forms respectively? and please furnish copies of every kind you use.

59. What rents do you average in the city and in the country?

60. How many hours per day do your clerks or operators work?

61. Do you employ female labourers, and if so, how and at what expense?

62. Have you ever paid damages by errors in messages? and has the responsibility ever been tried at law? and if so, please give the case.

63. Are you in the habit of sending free messages? and if so, to what extent? Also cypher messages, and what are your rules upon the subject?

64. Do you send news for the press at reduced rates?

65. Do you ever lose or mislay messages in their transmission ; and if so, does it occur often?

66. Do you ever pay back money on account of delayed messages?

67. Do you ever give any class of messages preference in any manner?

68. Do you require pre-payment ; and if so, on what kind of messages?

APPENDIX

69. Are the operators allowed to answer messages, giving information to a patron, at a distant office?

70. How many clerks are required to attend one instrument in a city or country office?

71. What system of telegraph do you use, and the cost of the apparatus? and, if possible, please give me the early history of its invention, and by whom? Please refer to any printed authorities, if any; and also to persons who are acquainted with any facts pertaining to its early history. Please give extracts, if you have any, from newspapers, magazines, or letters in your possession, pertaining to the above points, their date, and where they can be procured or examined.

72. Do you often make mistakes in messages; and if so, what causes the same?

73. Do you usually repeat back messages? and what are your rules respecting the sending or receiving of business on the line?

74. How many messages are you in the habit of sending before being advised of their proper reception from the office receiving?

75. How long has the plan of insurance been in use on your line, if at all; and is it any advantage to the Company or public? and if so, what is that benefit?

76. Do you insure on messages going beyond your line, and upon what plan? Please state the details, and give the forms adopted fully, whether going on your line or beyond, or from other lines.

77. Please give your opinions as to the use of magneto-electricity for telegraphing, and the expense of its application. How is it applied, and upon what length of circuit can it be employed?

78. Have you any mode of generating a continuous current of magneto-electricity; and do you think it could be continuously generated, giving an even or equal current, suitable for telegraphic purposes?

79. Do you work your wires charged continuously with electricity?

80. Are there any disadvantages arising from a continuous current, other than unfitness for your particular system; and if so, what are they?

81. What kind of submarine crossings do you consider the best, and how made, their cost, and by whom manufactured?

82. Do you consider there is any advantage in galvanizing the wires for cables, and to what extent?

83. Have you any facts relative to the extent of the action of the sea-water on the exterior wires? If so, please state them.

84. Do you know to what extent the sea-water acts upon the gutta percha? If any, please state the facts?

85. Do you consider there is any necessity for galvanizing the exterior wires for cables intended for fresh water crossings?

APPENDIX

86. Please give me all the information you can as to the early history of submarine crossings, with plans and principles?

87. Supposing you needed ten conducting wires, how would you advise a cable or cables to be made?

88. Do you consider a cable of more than six conducting wires practicable ; and if so, how constructed?

89. What is the weight of the cables of one, two, three, four and six wires, and the cost of each made of copper wire covered with one, two, or three coatings of gutta percha, being of No. 1, 2, 3, 4, and 5, as marked at the gutta percha factory, embracing the price of the respective materials?

90. Have you any information relative to the effect of lightning upon the submarine cables?

91. How do you protect cables from the dangers of lightning?

92. At what speed can a cable be manufactured?

93. Do you know of the use of gutta percha on lines overground, and how does it answer?

94. Supposing your line formed a circuit of two hundred miles, and there were fifty offices on that circuit, could you communicate with all the offices at one and the same time, and could they answer back respectively? And further, supposing one or more branch lines diverged from the main line at any one or more places, on which might be ten or more offices, can any one office on the main or branch lines communicate with all or any one of the offices on the main or branch line at the same time and at will, and be answered back at will? If so, by what arrangement?

95. Please state what were the first batteries used on the telegraphs in your country?

96. Please state what were the first telegraph lines erected in your country, how built, how long, when put up, when and how worked, by whom, and with what success? Also, what instruments were used on them?

97. Do you know any improvements in the art of telegraphing, either as to the lines or working, or as to the science not herein embraced, which would be beneficial to the enterprise if adopted?

98. Can you suggest any plan by which the telegraph can be made to serve the interest of the Government of the country relative to army, police, or other departments; and do you ever aid the police in the arrest of fugitives from justice?

99. Do bankers pay out money on messages from a distance, or delay protest ; and are your messages recognised as evidence in court between parties as to contracts ; and are your operators compelled by law to reveal in court the business of the line in any manner?

100. Can you give me any information relative to the early history and final invention of the different telegraphs? Please be particular,

APPENDIX

and give dates and the different stages of success, extracts from newspapers, magazines, books, etc., in which references are made to any or all of the inventions in question.

Here are Charles Bright's answers :—

FACTS PERTAINING TO THE SYSTEMS OF WORK AND GENERAL MANAGEMENT OF TELEGRAPHS IN ENGLAND, IRELAND, AND SCOTLAND.

BY CHARLES T. BRIGHT.

(Answers to Mr. Shaffner's Questions.)

Answer 1st.—Both. In its most important districts, from London to Birmingham, Manchester, Liverpool, Glasgow, Belfast and Dublin, the wires are laid underground. In some lengths there are duplicate lines, one above the other, underground. The following will show the extent of each description of telegraph in this Company's system :—

UNDERGROUND WIRES.—From London to Liverpool, by Birmingham, Manchester, Bolton, and Wigan, 250 miles, 10 wires. From Liverpool to Carlisle, 130 miles, 6 wires. Carlisle to Portpatrick, by Dumfries, 125 miles, 6 wires. Submarine cable from Portpatrick to Donaghadee (22 miles), 27 miles of cable used, 6 wires. From Donaghadee to Belfast, by Newtonards, 32 miles, 6 wires. From Belfast to Dublin, 105 miles, 6 wires. From Dumfries to Glasgow, and thence to Greenock, 115 miles, 6 wires. From Cork to Queenstown, 16 miles, 6 wires. Street work in London, Liverpool, Glasgow, Dublin, and other towns, 13 miles, 12 wires (average). On Scottish Central, Great Northern Railway, and Haigh Colliery Lines, 8 miles, 4 wires. Total, 821 miles of line—6,348 miles of wire.

Overground—Chiefly 6 wires.

On the Great Southern and Western Railway .	170 miles.
Midland Great Western Railway . . .	150 ,,
Dublin, Drogheda, and Belfast Junction and Ulster Railway Companies	160 ,,
Belfast and County Down Railway . . .	40 ,,
Belfast and Ballymena Railway	40 ,,
Ballymena and Coleraine Railway . . .	30 ,,
Londonderry and Coleraine Railway . . .	50 ,,
Londonderry and Enniskillen Railway . .	60 ,,
Kilkenny Railway Company	30 ,,
Waterford and Limerick Railway . . .	80 ,,
Caledonian Railway	200 ,,
East Lancashire Railway	100 ,,
Killarney Junction Railway	50 ,,
Portarlington and Tullamore	30 ,,

Miles	1,190
Wire	7,200

APPENDIX

The total mileage of the Company is therefore a little above 2,000 miles, and the length of wire about 13,000. The works in progress will bring the mileage to nearly 2,500 miles, and the length of wire to above 15,000 miles.

Answer 2nd.—TELEGRAPH POLES.—All the Magnetic Company's poles are larch. During the first seven years of pole telegraphs, (Cooke's patent for his mode of fixing wires on poles, the precursor of all other systems of poles in England, was dated September 8th, 1842, and specified in March, 1843), the timber used was, without any exception that I know of, Memel *squared* timber, chamfered down the sides. A table of the dimensions of these posts is given in Highton's book. Since the end of 1850, larch has been altogether used. All the companies have adopted the round wood in preference to the Baltic timber, from its being cheaper and more readily obtained, and if straight and well selected, stronger than the old wood.

We have no proof of the respective durability of the two woods, save from the comparison of gate-posts, etc., where the woods have been exposed, as in telegraph poles, to *wet and dry*, and we are led to consider that the larch poles will last much longer. None of the larch poles fixed have given way as yet, of course ; but most of the square poles fixed up to the beginning of '47, have become so much decayed immediately above and about the ground, as to make it necessary to lower them, which the height of the pole above the ground (14 feet) has generally allowed. The 4 feet buried in the ground being cut away, the pole is lowered to near the same depth. It must be borne in mind, when thinking of the safety of such short poles, that in England all the pole system is by the side of railways, and within their fence, and that persons who might injure the wires if fixed so low on the high road, have a wholesome dread of trespassing on a railway.

On a few lines where the poles have not been high enough to admit of their being thus lowered, they have been cut off at the ground, and fixed in a cast-iron screw socket—similar to the dwarf-screw piles used for breakwater fastenings, etc., patented by Mitchell.

I do not stipulate for any particular *age* of timber in purchasing larch poles, but only as regards the quality and dimensions. The age of the poles we use depends very much on the district we are passing through. In some parts the tops, and seven or eight feet of thick butt ends of poles, are used for sleepers, and for props in coal pits, and in others larch is only used for fencing ; and here we use the entire tree, from the butt to such part of the top as suits us, for size, while in the former case one pole would be the middle of much finer and older wood.

The size I fix is 18 feet in length, by 9 inches diameter, at the lower end, and 5½ to 6 at the top, measured after being barked.

APPENDIX

Crossing poles vary from 20 to 28 feet, according to the height of the railway cutting.

Answer 3rd.—I have the poles well charred, from the lower end to about a foot above the depth they will be fixed in the ground, and the charred part soaked in *gas tar* for about twelve hours, the poles standing in tanks of tar within a timber framing.

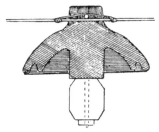

SECTIONAL VIEW OF BRIGHT'S INSULATOR

Answer 4th.—INSULATIONS. — The insulators first used by the Magnetic Company were of gutta percha, of the *surface* character of insulations, being simply two oblong pieces of gutta percha, about five inches in length, laid together while warm over the wire at the point of support, and fastened to the post by a small cast-iron chain, or shoe, screwed into the post.

These have been abandoned as not suitable to the long circuits the Company works, and a glazed earthenware insulation, of the roof character, substituted and adopted in all the Company's recent works.

I give a drawing to illustrate the insulations. You will see that the wires are so arranged upon the arms that any wire breaking will not fall upon the others.

This system of pole telegraph was patented by my brother and myself in October, 1852. The cost of each insulator, with bolt leaded in, nut and washer, is sixpence, delivered within two hundred miles. Of course, the apex of the cone in arranging the wires can be either above or below ; but I prefer its being below, as the wires, in falling, keep clear of any insulators below.

I find the insulators practically the best I have ever tried, though their size makes it necessary to have the poles well rammed.

It may be interesting to describe the various methods of insulation used in England in the order of their adoption.

The first, known as Cooke's Pole System, was by passing the wire through earthenware insulators of the size and form of an egg, slightly flattened at each end ; but the system, though simple itself, was hampered with a method of winding of the wires at each quarter mile, by means of ratchet wheels. The system is described in Walker's *Telegraph Manipulation.*

APPENDIX

Cooke's insulators were extensively used until 1848; but it was found that the surface was not sufficient, and an insulator, patented by Mr. Ricardo, but generally known as Physick's insulator, was brought into use. It is described in the *Mechanics' Magazine* for 1850. The wire is supported by a hook, the upper part of which passes through a shed of earthenware, and is fastened by a nut at the top; above this mastic was laid, to insulate the hook from the post. This was found a very faulty insulator, the vibration of the wires, and other causes, breaking off the mastic.

The next insulator is that known as Clark's insulator, having been patented by Mr. Edwin Clark, the engineer of the Electric Telegraph Company, in 1850. It is described in one of the numbers of the *Repertory of Patent Inventions*, 1851.

This insulator is now being discarded for one of the same form, and

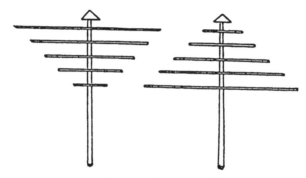

BRIGHT & BRIGHT'S PATENT, 1852

hanging down from an arm, but made of glass throughout, without any metal cap.

The object of the metal cap was that the moisture might rather condense on it than on the earthenware; but it always seemed to me very hazardous to have a band of metal of such surface so near to the earthenware, and the result shows that the principle is faulty in practice; for the pent-house formed between the metal and the earthenware becomes so clammy with dew, or fog—without rain, which, of course, adds to it—that the insulator is a very defective one.

The British Company simply adopted a plan of lapping the wire with silk ribbon for about six inches on either side of the point of support, and covering about five inches in the centre of the foot of ribbon with a piece of gutta percha, shaped like an elongated sphere; the whole is then varnished with brown hard varnish. I believe this is very fair insulation for a few months, but the varnish soon comes

427

APPENDIX

off, unless frequently renewed, which is very expensive, and the silk decays and holds the moisture.

With Clark's and Physick's roof insulators the closeness of the sides was objectionable, on account of the clamminess thereby engendered, and the difficulty of cleaning.

In our roof insulator we have spread the sides out, so as to widen the dry surface, and, by making more surface for the air to dry the insulator, the damp and dew is sooner dispelled. There is less lodgment for insects, and the insulator is readily cleaned. The double channel round the wire is to prevent the rain being blown up the insulator in high ground. I have adopted the method of fixing the insulator from below instead of from above, as I find it firmer, and there is no strain on the fixing of the bolt.

Other insulators beside those I mention have been invented ; indeed nearly every patentee has had an insulator among his claims, but none of them have ever been used, except Brett and Little's, Nott's and Bain's, all of which were removed very shortly after their establishment on short lines.

One of the most extraordinary ideas on the subject of insulation is that of Highton, who in his patent, dated January, 1852, proposes to run a wire down each post to the earth, from the central point between each pair of wires, so that any of the " electricity transmitted, as it escapes from the wire, may be intercepted by this communication with the earth, and so transmitted direct to the earth without the possibility of its entering an adjacent wire " ! I have no doubt a wet day would satisfactorily prove that more than he wished would be intercepted. It has not been adopted.

Answer 5*th*.—STRUCTURE.—Poles of the dimensions above mentioned cost 3*s*. 6*d*. or 4*s*. each, barked, the knots planed off smooth, and the lower ends charred and tarred.

Twenty-five are generally fixed per mile (unless there are other supports, as walls, buildings, bridges, or viaducts); the number used to be 30, and frequently 32, but it is now preferred to strengthen the poles, and sink them deeper, and by fixing only 25, to reduce the number of points of suspension for the wire, and thereby improve the insulation.

The expense of erecting the line in labour varies very much according to the price of wages in the district, the nature of the soil, the fitness of the weather, and the length of the days at the time, etc. ; but it may be taken from £3 to £5 per mile. I am estimating for six wires.

Answers 6*th and* 7*th*.—ON WIRES.—Galvanized iron-wire, No. 8 (Birmingham wire gauge), of the quality known as "best annealed." Cost, at present time, £23 10*s*. per ton, delivered within 200 miles, less 3 per cent discount for cash. Weight, 3 cwt. 1 qr.

APPENDIX

18 lbs. per mile. Manufactured chiefly at Birmingham, London, and Liverpool.

I think the difficulty and cost of keeping ungalvanized wire, properly coated with paint to prevent oxidation, renders it unsuitable for telegraphic purposes, otherwise it would be stronger, especially at the welds, than galvanized wire. Near large towns there are disadvantages in galvanized wire.

No. 8 has been universally erected in England, except on the length from London to Southampton, one of the earliest lines built, when No. 7 was adopted. It did not appear so much broken by frost, etc., as to influence its being selected afterwards.

In very important circuits, I use stronger and more expensive wire, known as "best charcoal annealed," which is sold at the present time at £29 per ton.

The cost of wire has been gradually rising for a long time. In 1851, best annealed was quoted at £16, in 1852 it had reached £18 10s., and it has advanced steadily since that time. The wire is generally delivered in quarter-mile bundles.

GAUGE OF WIRE.—I do not know if your numbers of gauge are the same as ours ; perhaps the following table of our *lengths* of *one pound* may be a guide to you if they differ, in understanding the size— I mean by any number of gauge I speak of.

	ft.	in.		ft.	in.
No. 1.	4	0	No. 13.	41	0
4.	6	8	14.	55	0
6.	7	3	15.	66	0
8.	13	6	17.	113	0
10.	21	6	18.	150	0
11.	28	0	19.	206	0
12.	33	4	20.	250	0

These figures are from personal weighing and measuring.

Answer 8th.—MODE OF JOINTING.—I have all our joints soldered. Those made on the line are as in sketch below.

The ends of the wire to be joined are bent about ½ an inch ; being laid together, they are lapped or bound with galvanized iron-binding wire (No. 20, B. W. G.) and soldered. I do not know any telegraph wires breaking at the *joints* made after this fashion.

The wire used to be welded at the manufactory, and the weld more frequently gave way than any other part of the wire. Latterly, within the last two years, we have adopted a different plan—similar, I believe,

APPENDIX

to that used on many of the American lines. The two ends are laid side by side for about 5 inches, and each lapped four or five times round the other, with a space between each helix of about ¾ of an inch.

Answer 9th—ATMOSPHERIC ELECTRICITY, ETC.—I have had no opportunity of observing any difference between galvanized and ungalvanized iron wire in their electrical statics during atmospheric variations ; but I have no reason to suppose that there would be any difference—at all events, not so as to be perceptible on ordinary telegraphic apparatus.

We suffer most from atmospheric deflections in April, and the end of October and November. We have not arrived at any laws, nor does it seem, from the irregularity of the occurrence of these magnetic variations, that any definite laws can be established for their appearance or extent.

Some interesting experiments carried on for some time on a pair of spare wires between Derby and Birmingham, in 1848, are given in a paper read by Mr. Peter Barlow before the Royal Society in 1848, where the variations are tabulated for many days.

The term "deflection," which we use in speaking of the varying currents induced by magnetic storms, arises from Cooke and Wheatstone's needle telegraph, being at first so generally used in England, and the needle being *deflected* for awhile to one side or the other by the temporary continuousness of the atmospheric currents. The effects appear to be the same in this country and on the Continent as in America—at least, I have not heard of any more violent or frequent deflections in America than in England.

Answer 10th.—In wet weather or long circuits we sometimes experience difficulty from currents passing from wire to wire at the poles, depending, of course, in extent on the degree of good or bad insulation of the line.

Answer 11th.—No.

Answer 12th, 13th, 14th, and 15th.—Yes ; the coils and other parts of our apparatus are sometimes fused and broken, and the needle demagnetized, but I have never heard of any property being burnt or injured through the agency of telegraph wires. I should expect that they have rather been the means of saving considerable damage by carrying off a great deal of electrical matter. I have occasionally had poles injured by lightning, but cannot supply any certain scale of average, not having kept record of each case.

I have had only two cases since the commencement of this year, in

APPENDIX

above a thousand miles of pole line. One of them was the most violent that has occurred in this country, twenty-four poles being more or less injured near Newry, on the Dublin and Belfast line, eight being split open and splintered to their bases, so as to be totally useless again.

A wire conductor, terminating in a spike, used to be let into all the squared poles ; but since larch round poles have been introduced, it has been abolished.

We have violent electrical discharges, as often with as without thunder—of course more on long circuits.

We use lightning protectors with our instruments on pole lines. We have hitherto used two plates with saw-teeth screwed into a mahogany base, the points being so close as nearly to touch—one is connected to earth and the other to line, and a small spiral, all of very fine copper wire (No. 40), forms part of the line circuit. The whole is covered with a glass cover. It is a very simple arrangement, and effective, but it is found objectionable on account of the wood sometimes warping and bringing the earth in contact with the line.

I am about to introduce a protector, included in Bright & Bright's patent of 1852, formed of two wire brushes (about the size of a nail brush, the wires being about $\frac{1}{8}$ of an inch in length), mounted on plates of brass $\frac{1}{2}$ an inch thick—the top plate being capable of ready adjustment by a screw.

I have lately seen in Turnbull's work on the telegraph in America, that something after the same plan is in use there, and known as Carey's Protector, and I am curious to know the date of its introduction. I can easily understand the difficulty in adjustment, if the points are mounted in leather, as Carey's protectors are thus described.

I am under the impression that in your country lightning is far more frequent and injurious, and more violent than here. Perhaps this may partly arise from the wires passing through districts less populated, where there are not the same number of conductors in works, chimneys, towns, railways, etc.—and I should like to know if you find the same average of discharges in your thickly populated districts as in those where the towns are distant from each other. Say from Boston to New York, and Baltimore and Washington, compared with some long line in a thin district.

Answer 16th.—Answered under questions 2 and 5.

Answer 17th.—The size of poles I have described is intended for not more than 8 wires. *We* have no greater number anywhere on poles, except at a few junctions where different lines meet and pursue the same course for a short distance. They would probably bear 10, but I should not like to have so many in an exposed country without increasing the size. The greatest number of wires that I am acquainted with on one set of poles for any distance is on the East

APPENDIX

Counties Railway, where 18 wires are carried from London to Stratford. The London and North Western line has 13 wires, the greater part of the distance on one set of poles, and in many places for short distances many more.

Answer 18th.—I have not observed any marked difference arising simply from position of the wires. I should suppose the lower wires would be more affected by earth contact, and the middle wires with wire contact, while the top ones would be the best ; but to ascertain this it would be necessary to have all the wires exactly on a par as regards their insulation, and a delicate galvanometer would scarcely then show much difference. I should imagine ordinary instruments would indicate no appreciable difference in their working, if the wires were otherwise equally insulated.

Answer 19th.—In our old method of insulation, the gutta percha held the wire at each pole. In our present plan, you will observe the pin, which passes over the wire across the slot in which the wire lays, has a piece of binding wire laid over it, and lapped and soldered around the wire, on both sides, so that, while the wire has a little play at the point of support, it is held in case of breakage. If a wire is held by anything biting fast to the wire, the continual oscillation at that point injures the fibre and makes it very liable to give way there.

Answer 20th.—Not often after the first winter, which of course tries them more severely than any time after. Contraction by frost, weight of ice and snow, and accidental or malicious injury, are the general causes. I have known galvanized wires near manufacturing towns broken by becoming so attenuated in some places, in consequence of the destruction of the zinc by gradual deposit and decay, as to break by their own weight.

Answer 21st.—By line men kept on the line for the purpose, and for general maintenance of the poles and wires. The number employed varies on different lengths, according to the importance of the line as a commercial circuit, and consequent necessity or otherwise for immediate repair of any fault, and according to the convenience for speedily getting to any place in the length. On a straight line of railway, for instance, any point of a man's district may be much more readily arrived at than in one composed of many short branches, where the junctions and changes of trains make it difficult to travel so quickly. The average on my lines is one to every 70 miles of pole line.

Answer 22nd.—No.

Answer 23rd.—Answered under 21. The men's wages are from 18s. to 24s. per week. The latter for men of intelligence, capable also of repairing the instruments, or undertaking charge of works to small extent.

Answers 24th and 25th.—Our winters are not so severe, and our

APPENDIX

frosts so long in duration as to give us very much inconvenience in this way. Sometimes the wires break from the weight of the snow, when it collects to a great extent between the poles, but it is a rare cause of interruption. The greatest injury I remember being done by snow was on the South Eastern line, where a considerable distance, above two miles I think, of poles and wires were thrown down.

Answer 26th.—SUBTERRANEAN LINES.—We have 820 miles of underground line, 670 of which, from London to Dublin, by Manchester, Liverpool, Carlisle, and Belfast, are in a continuous line, the longest underground line by far in the world.

The chief part of this is laid in a trough of kreosoted Baltic timber, with a lid of galvanized roof iron, overlapping the groove by $\frac{1}{2}$ an inch on each side, of the gauge No. 14 in thickness.

It is drawn with six wires, but in some places ten are laid.

The line from Manchester to London, the first laid, has a wooden lid instead of the iron lid afterwards introduced. The district is easy of access by railway the entire distance, and the roads well attended to by the road surveyors (country, not *telegraph* officers), who inform us of any works, etc., to be done on the line of our wires.

The wires on this line, ten in number, are covered with a serving of tarred jute as an additional protection, especially while laying, the expense being nearly covered by the saving in labour and carriage, in having the wires altogether in a rope, and wound on the same drum.

The two plans are under the ordinary high road ; but through the paved streets of towns, where the roads are often opened for laying gas and water pipes, drains, etc., and where, from the nature of the ground, the full depth of the trench cannot be made, the wires are laid in cast iron pipes.

The proportion of street work is generally about 3 miles out of every 100, but on some lines considerably more ; between London and Manchester we have 21$\frac{1}{2}$ miles laid in iron pipes out of 200.

Street wires used to be drawn through solid gas piping of about 3 inches diameter, the pipes being laid first, and the insulated wires drawn through afterwards. In doing this the insulating material was frequently injured ; sometimes the wires were broken inside the gutta percha or other insulating material by the force necessary to pull them through, and occasionally they were drawn so tight that on the slight settlement of the ground, usual after the line has been laid a short time, some of the wires broke inside the insulating material, occasioning great difficulty and expense in detecting the fault.

The great proportion of the faults, however, were only abrasions of the insulating material ; and though at the time the wires passed with all appearance of perfection through the ordeal of testing, and the streets were closed, and the pavements reinstated, before long the

APPENDIX

defects became so manifest as to interfere with the working of the apparatus, and the streets had to be re-opened, and the wires tested through length by length for the fault.

The wires required jointing at every other drawing point, and these points frequently proved defective, particularly in the old varnished cotton method of insulation and others prior to the use of gutta percha.

In the beginning of 1852, having considerable lengths of street work to lay, I gave a good deal of attention to the subject, and determined on having the pipes cast longitudinally in two pieces, so that the wires could be *laid in* the under lengths, and the upper lengths then attached, instead of drawing or threading them through solid pipes. I was the better able to carry this out, through the introduction of gutta percha, rendering the exclusion of moisture for the interior of the pipes of less moment. I tried various forms; rectangular, half-rectangular, with an arched lid, semi-cylindrical, with a flat sole, etc., but the form I found most generally useful and convenient was that having the upper and under half exactly similar, making together a round pipe. I have the pipes cast in 6 foot lengths, and about 2 inches internal diameter, the substance being $\frac{3}{8}$ of an inch—the sides fitting clean together, without any flange, but fixed by small bolt and nut fastenings through semi-circular lugs projecting about $1\frac{1}{2}$ inches from the side ; one pair of lugs being about 9 inches from the faucet, and another pair 2 feet from the spigot end.

A pipe of these dimensions is much cheaper than the old 3 inch solid pipe, and more generally useful—the halves being convenient for fixing to walls, viaducts, etc., over wires needing good protection in such places ; and, from its circular form and smallness, it is very difficult to break, as a pick-axe, or other tool, cannot easily strike it full.

The process of laying in the wires is rendered much more expeditious and economical by the use of half pipes. I select a chronicle of some speedy operations in the middle of 1852, from the *Times* newspaper :—

"There is no greater annoyance in large towns than that which Parliament has granted to private companies of ripping up their pavements at all times and places where it may be necessary for their interests. The authorities of a town pave their streets at great expense, and then comes a gas company, and then a water company, and then a telegraph company, to open deep trenches in some of the leading thoroughfares, interrupting the traffic, and creating great inconvenience. We have seen these trenches open in some cases six or eight days ; but in the present instances we are glad to perceive that a great improvement has been introduced. The method adopted by the old Company has been to lay down a line of round

434

cast metal pipes, through which the insulated wires are passed. This is necessarily a long and tedious operation, because considerable time is occupied, as each length of pipe is laid down, in passing the wire through it ; but Mr. Charles Bright's plan is to use pipes split longitudinally into two halves. The under halves of the pipes are laid down in the trench, and then a large drum, on which the insulated wires are wrapped, is rolled along over the trench, and the wire is payed off easily and rapidly into its place—the upper parts of the pipes put on afterwards, and secured in their places by means of screws through small flanges left outside for the purpose.

"So well has this mode succeeded, that in Liverpool the whole lengths of the streets, from Tithebarn Railway Station to the office in Exchange Street East, were laid down in a single night (11 hours), and in Manchester, the line of streets from the Railway Station in Salford to Ducie Street, by the Manchester Exchange, in 22 hours. This was the whole time occupied in opening the trenches, laying down the telegraph wires, and re-laying the pavement ; and while great credit is due to the Company on the ground of the little public inconvenience occasioned, no doubt they would find the benefit of it in economy of time and money."

Mr. Reid has invented an ingenious modification of the half pipe, of the rectangular form, which he has patented in company with Mr. Brett, and which we have used. I refer you for this to his patent, published in *The Repertory of Patent Inventions*. Mr. Henley also has improved on the circular half pipe where it is intended only for subterranean work, which he has also patented ; but both of them have top and under lengths differently shaped, and I find my original plan preferable for general purposes. All the telegraph companies have adopted the two-piece pipe in place of the solid round pipe, except the old Company. The depth of our trench is not less than 2 feet, but all obstacles, as drains, culverts, gas or water pipes, etc., are always passed *under*.

Answer 27th.—I have had no experience in laying underground wires with single covered gutta percha, having in common with all telegraphic engineers in this country considered the occasional small flaws and air bubbles which occur in single wire, and which are covered and made good by the second coating, a bar to its use, except about stations, etc., where it is not in close contact with the earth, and may be readily examined.

Answer 28th.—I do not think wire, covered with hemp only, could ever be laid so as to preserve good insulation, equally with that coated properly with gutta percha.

The wires through the streets of towns used, prior to the introduction of gutta percha, to be coated with a double serving of cotton, varnished, tarred, and enclosed in a leaden tube, which was passed

APPENDIX

through cast iron 3-inch piping. The wires were continually getting defective after being laid some little time, and we have only been able to have underground wires of any length in a good state of insulation since the adoption of gutta percha, and that only within the last five years. Before that, the art of coating wires had not reached its present high state of practice (which may be attributed to the perseverance, energy, and science of Mr. Statham, the able manager of the Gutta Percha Company); and in one of its first trials in the most important lengths of street wires in London, it proved in a few months to be an utter failure.

Answer 31*st.*—Answered partly under 26. The cost of laying varies very much according to the hardness of the roads, the price of labour, the season at which the work is done, etc.; for six wires, according to the plan shown in sketch appended to Question 26, a line along the old mail-road varies from £180 to £200. The price of gutta percha has changed so much as to make estimates very little to be depended on for a long time. Last year, No. 4 rose £2 per mile in three months, and other gauges in proportion. For ten wires, according to the plan with wooden lid, and covered with hemp, the cost may be set down at about £230 per mile—this is on hard macadamised roads.

I should never lay less than four wires underground; the proportionate expense of cutting the trench, and for troughing, etc., being about the same for one as for ten, unless the scarcity of timber be much reduced, the expediency of which I doubt.

Wires laid without some protection cannot be depended on very long, unless in a very favourable country. In Prussia they appear to have formed the same opinion. We have had to re-lay a line from Manchester to Liverpool, which was originally laid without protection, though sunk to a good depth. A line of two wires laid from Dumfries to Stranrae, in Wigtonshire, by a now defunct company, called the Channel Submarine Company, has never been worked, and never will be.

The depth of our trench is two feet. In towns, and where gas and water pipes, etc., are laid, more, according to the level of the mains and service pipes, which we keep under in all cases.

The only other company which have a line underground of any length (the European and Submarine, from London to Manchester), have laid their wires less deep—from one foot to eighteen inches.

Answer 32*nd.*—Where the road is rocky, we blast out about a foot deep, and lay the wires in iron pipes, packing up the trench with the shale and earth. We have had a great deal of rock crossing Shap Fell; on the road from Liverpool to Carlisle, we had a considerable length of solid rock; on the London line about Stoney Stratford; on that from Dumfries to Glasgow, near Abington; and through the Deloin Pass, and a good deal in Ireland.

APPENDIX

Answer 33rd.—Our wires are in every case, as yet, laid along the old mail-roads, which have been so carefully made and kept in repair throughout the kingdom for years past ; we do not therefore ever pass through *marshes*, as the road would always pass over anything of the sort with a bridge or viaduct. We have no telegraphs in England "across country" without regard to roads. For the same reason, we have no upheaving of the roads from frost ; they are all too old and firmly set for any such disturbance. The only danger at all of the sort that I apprehend is the *settling* of the roads in some places in the colliery districts, from seams of coal mines passing under the roads.

Answer 34th.—If you mean underground wires, at present, I cannot say which. I should imagine that seasons do not affect underground wires, save in induced currents, at all ; with pole lines, spring, autumn, and winter, are the worst times. I think about in the above order ; summer is of course the best season for exposed wires.

Answer 36th.—See reply to 33—Our mail-roads always cross by bridges, and our wires are laid over them, frequently close under the parapet, about 6 inches deep (as the crown of the bridge is generally shallow, to avoid much rise of the level of the road), enclosed in wrought iron solid pipes, about an inch in diameter, by three-sixteenths in substance, which are threaded over the wires, for the short distance required.

Answer 37th. — Cross, Induced, and Return Currents. — Under question 10 you have referred to *cross currents* as what we term "wire contact," that is, communicated by moisture, or otherwise, from wire to wire.

Under 11, you speak of an *induced* current from wire to wire.

I suppose, therefore, that by the latter you mean something that we have not yet experienced ; at least, not so as to be visible on our ordinary instruments, but which perhaps your continual current, or some peculiarity of your atmosphere, in some districts may have engendered.

We have had no case of wire contact in underground wires ; I should doubt our having any, unless in some cases of a nail, or a portion of an iron cover, or of a tool being forced in between two wires, as the communication with the earth (or to use our term, the "earth contact") would be paramount.

We have had no experience of any induced currents from wire to wire in underground, any more than on pole lines.

Answer 38th.—I have not observed any difference ; certainly our instruments are not in the least affected. You must bear in mind that our wires have been laid only a short time. That from London to Liverpool has been completed nine months—the North and Irish lines only six ; and having been very much occupied with extensions,

APPENDIX

and opening out our present system, I have had little time for experiments with delicate galvanometers and other apparatus, but I speak here, and before, of my results as shown on working telegraph instruments.

Answer 39th.—I cannot speak for certain, as I have not had time to try. I have in reserve this, with a number of other experiments, when I have more leisure, and will acquaint you with the result.

You are aware that we use magnets only, and their induced currents, for our motive power ; and as my experience of batteries has of late been only for purposes of testing, etc., this and the two following questions will probably be much better answered by your other friends here.

Answer 42nd.—Answered under 12, etc.

Answer 43rd.—On overground lines they are very trifling indeed compared with underground ; the conditions on which the wires are suspended and insulated, passing also through a medium, capable to a certain extent of absorbing any electricity developed in surplus, prevents the occurrence of any effects appreciable by ordinary needle telegraphic instruments.

I look upon an underground wire as being exactly similar, on a large scale, to a Leyden jar, and I am borne out in this by the experiments of my brother and myself, and by those instituted by Faraday on the underground wires more recently laid by the Electric Telegraph Company. The magneto-electricity, as well as the galvanic (or chemical) electricity, evinces these phenomena, hitherto supposed to belong to properties appertaining peculiarly to frictional electricity,

The copper may be compared to the inner metallic coatings of a Leyden battery, the gutta percha to the glass, and the earth and moisture surrounding to the outer covering.

I was much interested in one of our experiments to observe that the larger the size of the wire experimented upon, with the same battery power, the greater the amount of return current : a strong support of our opinion ; as, had it arisen from an *elastic* return, owing to the wire being unable to receive as much electricity as was forced into it, as some supposed, of course a *smaller* wire (with the same power as that employed with the larger size), should have given out a *greater* amount of return current. If you experimentalize on No. 18 and No. 16, you will see this very clearly.

Answer 44th.—My brother tried some experiments, by connecting our underground wires together, which he will be better able to describe than myself. I am about to try, on a very extended scale (over all our wires joined together, and various lengths), the distance to which we *can get a current*, and will acquaint you with the result.

Answer 46th.—Practically the difference is that an underground line of copper requires considerably more power than a suspended

APPENDIX

line of iron. If we assume the difference of conducting power of the two metals to be in the proportion stated by Becquerel and others, then the vast resistance engendered by the (so to speak) Leyden jar condition of a well-insulated underground wire is manifest.

Answer 45th.—No. 8 has been the only size of wire used overground. No. 7 has been tried on one line, but No. 8 was found to be sufficiently strong. No doubt, a much smaller wire would have sufficed (in the absence of any appreciable return current, as in subterranean lines) for all ordinary conducting powers.

I have tried with Mr. Statham, the Manager of the Gutta Percha Company, some important experiments on the conducting ratio of different sizes of gutta-percha-coated wire, which show a considerable difference in conducting power of copper wire, leading me to the decision that, for any line of a length above 100 miles, it is not expedient to use a size less than No. 16 copper wire (Birmingham wire gauge).

I extract a few notes of our experiments on Nos. 18 and 16, as clearly showing the difference.

50 miles of No. 16, and 50 miles of No. 18, tested for continuity with a galvanometer.

	50 miles of No. 18.	50 miles of No. 16.
(1st) With 3 pair of plates,	45°	. 53°
,, 6 ,, ,,	62½° full, so as not to be reckoned in degrees.	
	100 miles of No. 18.	50 miles of No. 16.
(2nd) With 3 pair of plates .	29°	. 39°
,, 6 ,, ,,	. 50°	. 59°

With needle of galvanometer weighted.

	100 miles of No. 16.	65 miles of No. 18.
(3rd) With 36 pair of plates	17¼°	. 17¼°
,, 18 ,, ,,	. 10¼°	. 10¼°

The loss of conducting power appears, therefore, to be more than proportionate to the different area of metal.

Mr. Statham, who went to considerable pains in investigating the ratio of conductibility, and insulated many miles of various sizes, practically to demonstrate the differences, has made a very good wire for short distances, by which a saving may be effected in such lengths of underground lines, which he calls No. 7 gutta-percha-covered wire, the copper wire being 18—the gutta percha being *double* covering, but not so thick.

Answer 77th.—MAGNETO-ELECTRICITY.—I am naturally strongly in favour of the use of magneto-electricity. Its economy is undoubtedly the most prominent feature. A pair of magnets, costing at Sheffield 30s. and perhaps 40s. to 45s. (according to the finish bestowed on the instrument), by the time they are fixed and ready for use, will send a strong current on a well insulated suspended line for above 200 miles, and on an underground wire above 100. (I have had signals, but only weak ones, through 250 miles of underground wires with the class of instruments I am speaking of), while the six twelve-cell trough battery used in this country, which would be necessary to perform the same work, would cost £7 10s., besides the constant expense of renewal, etc.

A magnet, if the keepers are put on when the instrument is not in use, will retain its magnetism for an indefinite time. I have not as yet had to re-magnetise any of our sending magnets, though we have instruments that have been in use since our incorporation in 1852, and some since the exhibition in 1851, where our apparatus carried off the highest medal the jury could allot.

I cannot speak as to what length of circuit magneto-electricity can be used, as I have not as yet tried with magnets of great power, beyond those we use daily.

We work the longest circuit that is daily worked in England—from Liverpool to Belfast and Dublin, an underground line, which, with a pole line would be equal, at least, to 800 miles, though I doubt if we could put a current of such power into any pole line, unless in very dry weather. The magnets here used are the large horse-shoe compound magnets you have seen in our offices, about 15 inches from the poles to the back, about 5 inches in height, made of 12 plates, in breadth about $1\frac{1}{2}$ inches. I have spoken with these for experiment through 530 miles of underground wire; but for our arrangement of circuits it would be unnecessary and inconvenient, as we should have to lay more wires so as to connect up long direct circuits, which would be very little used.

These magnets cost nearly £7 each.

Answer 78th.—We have no mode in use, because it is not adapted to our system, the magnetic current we send being sufficiently long for our purposes.

I do not see much difficulty in keeping up a continuous current. Wheatstone patented one some years since, but it was complicated and comprised six compound magnets for each wire.

A simple reverser, which should change exactly as the coils changed their polarity, similar to that used by Billant, and a close arrangement of the poles of the permanent magnet, are the chief points.

By rotating a disc or plate, with coils on its axis between the poles, perhaps an equal current might be maintained.

I have not given much attention to the subject, as we do not use continuous currents here, and can obtain sufficient duration of the effects of the movement of a pair of coils for about 30° in the face of a permanent magnet, by having the receiving coils and their cores so arranged as to retain the residual magnetism until another current is sent.

APPENDIX

Answer 79*th*.—We do not use our wires continuously charged with electricity.

Answer 80*th*.—I do not know of any practically, from having had no experience at all in the use of continuous currents. Glancing at the subject, I should think there must be some inconvenience wherever an iron core is used in the receiving coils, by the residual magnetism retaining the keeper or magnet. This would be increased in an underground line, where the return current would follow so quickly on any break of current, as to make any rapid sending impossible. Is not the expense of maintaining battery much increased by their more frequent operation?

Answer 81*st*.—SUBMARINE TELEGRAPHS.—I like the plan most generally adopted the best—that of covering the gutta percha wires first with tarred hemp, and afterwards with strong iron wires wound spirally round the rope. They have been made in England chiefly by Newall, of Gateshead. The cost varies very much of course, according to the size of wires used, etc.

Answer 82*nd*.—I do not think there is. It weakens the rope, and in manufacturing is liable to injure the gutta percha wires by scaling, etc. We lost a galvanized rope between Portpatrick and Donaghadee in October, 1852, which was entirely recovered afterwards. A new one, ungalvanized, was laid in May, 1853—the first successful attempt to connect England to Ireland after three failures by other companies and ourselves.

Answers 83*rd to* 92*nd*.—I have either been unable to give reliable information on the subjects, or cannot just now go into them fully, and collect estimates. I shall be happy to do so hereafter, if you have not obtained full particulars from others.

Answer 93*rd*.—INSULATED WIRES ON POLES.—Gutta percha wire has not been used in England upon poles. For my own part, I consider it would have all the disadvantages of both the overground and underground systems, with few of the merits of either. I speak of our abandoning gutta percha pole insulators under No 4.

Answer 94*th*.—COMBINING CIRCUITS.—We have no circuit with anything like that number of stations, so I cannot speak by experience; but I have no doubt if our present instruments could not overcome so great a resistance as that number of receiving coils in a long circuit would present, that apparatus could, without difficulty, be made that would be sufficient. We have no arrangement exactly such as you mention; but at all our junctions we have switches for turning the branches into direct communication with the main lines.

Answer 95*th*.—GALVANIC BATTERIES.—Trough batteries after Wollaston and Cruikshank's plans. Mr. Cooke modified them, and introduced the use of sand in the cells, which equalized the action, and made the batteries more convenient for carriage and use in offices. I speak of the first working telegraph—not experiments.

APPENDIX

Answers 96th and 97th.—FIRST ENGLISH TELEGRAPH.—Cooke and Wheatstone's Needle Telegraph. That with 5 needles, used on one or two lines at first—(the Norwich and Slough lines the only ones I remember at present)—but were shortly afterwards changed for the double needle as more convenient and economical. The first line—the Great Western to Slough—was fixed in 1839. It was erected on short standards, the wires being laid in a trough carried on them. It was worked a long time, but has since been changed for the suspended wire system. Its chief use was for the railway company's affairs, and partly for commercial business. Other lines followed rapidly after this—among the foremost, the South Western to Southampton, Gosport, Portsmouth, etc., with wires for Government use in the Admiralty Department.

DIRECTIONS FOR INSULATING JOINTS IN GUTTA PERCHA COVERED ELECTRIC TELEGRAPH WIRE.

HAVE in readiness a few strips, about $\frac{3}{8}$ inch broad, of very thin gutta percha sheet, also a little *warm* gutta percha about $\frac{1}{8}$ inch thick, one or two hot tools, and a spirit lamp.

Remove the gutta percha covering from along the wire no further than may be necessary for making the joint in the wire. Having joined the wire, warm gently with the spirit lamp the bare wire and point, and the gutta percha near to it ; taper the gutta percha over the bare wire until the ends meet ; warm this, and immediately apply one of the strips of thin sheet in a spiral direction over it. Press this covering well on until cool ; then, with the spirit lamp, carefully warm the *surface*, and proceed as before to put on a second strip of the thin sheet, observing to wrap it in a direction reverse from the last strip, always making the commencement and termination of these coverings to overwrap the previous one. *It is safer to perform this operation a third time.*

Next, take a piece of the warm $\frac{1}{8}$ inch sheet, and cover over the coats of thin sheet, again overwrapping the original covering of gutta percha, which should be heated so as to ensure perfect adhesion. Press it well on as it cools, and when cold, or nearly so, finish off the joint with a warm tool, working well together the old and new material at each end.

Lastly, and in general, avoid moisture, grease, or dirt, and be careful not to burn the gutta percha, which would prevent proper adhesion.

Charles Bright left a number of Colonel Shaffner's questions for his brother to answer, some of them as more immediately con-

APPENDIX

cerning the administrative and commercial aspect of telegraphy, and others as being connected with investigations in which the latter had been more intimately associated. The answers to these examining questions are appended here :—

THE SCIENCE OF TELEGRAPHING—MAGNETO-ELEC-TRICITY—RETURN CURRENTS—ROYAL PROTECTION TO TELEGRAPHS—BUSINESS DEPARTMENTS AND OFFICE ARRANGEMENTS.

By Edward B. Bright,
Manager of the English and Irish Magnetic Telegraph Company.

(Answers to Mr. Shaffner's Questions.)

Answer 44th.—Return Currents on Subterranean Lines.—In the course of a long series of experiments carried on last year by my brother and myself, inquiries were instituted with reference to the speed with which the galvanic or magnetic sensation is communicated through underground wires.

The result of the inquiry shows decidedly that the communication of the electric impulse through a length of 500 miles of underground gutta percha covered copper wire (16 gauge) does not exceed 900 to 1000 miles per second—a speed far below that usually assigned.

Reasoning upon the issue of these experiments, and those previously tried in America, I have no doubt that the speed of any description of electricity varies greatly with the peculiar conditions and nature of the conductor used, and also with the length of the conductor interposed ; and that a wire suspended in the open air, especially if insulated only at points of its support (such as in a pole line) would offer far less resistance (*cæteris paribus*) than a wire underground.

Submarine cables are similar, as regards electrical conditions, to subterranean lines, and the speed with which the electric impulse is communicated would be the same.

I have no doubt, however, that, as Professor Wheatstone's experiments showed, the speed of electricity developed by friction (or machine electricity) is of *far* greater velocity, whatever be the medium of communication interposed, than the species (voltaic or magnetic) used in telegraphic manipulation.

Answer 48th.—Preferred Messages.—It is provided in the Royal Charter granted to the Company that any communications handed to the Company by Government for transmission are to be forwarded prior to any other messages, and at all times when the offices of the Company are open. The Government make considerable use of the magnetic telegraph, and their messages always take precedence, and are paid for according to the usual tariff for private messages.

APPENDIX

Answer 49*th*.—RIGHTS OF WAY FOR TELEGRAPHS.—Government has made. no grant as regards concession of property to telegraph companies, but gives to them power to open roads, and to pass their wires through or over any public way or public property without compensation, and subject only to certain formal notices to the local road surveyors, and to their approval of the subsequent re-instation of the surface of the road opened.

The various telegraph companies have availed themselves more or less of the power so conferred,—the Electric Telegraph Company by laying their wires through the streets of towns from the termini or stations of railways ; this Company for some years past have extended the use of the clause to the construction of main lines along the high road, in addition to street work ; and recently the European Company have adopted a like plan. Finally, the British Telegraph Company, in several instances, have carried pole lines for short distances (a few miles) over the highway, where they could not obtain permission to use the railways. There are many difficulties, however, in carrying wires along the turnpike roads overground, as the landlords of adjacent property have particular objections to the lopping of any branches that project, as is often the case for miles on both sides ; and wires so placed are peculiarly open to malicious injury, which could not well be guarded against.

ROYAL PROTECTION OF TELEGRAPHS.—Accidental or malicious injury is provided against in the Company's Act of Parliament, dated 1st August, 1851, 14, 15 Vict., cap. cxviii. I cite the clauses that particularly refer to this point :—

" LIV.—That if any person shall wilfully remove, destroy, or damage any electric telegraph, which shall or may have been lawfully erected, or any wire, standard apparatus, or other part of such telegraph, or any works connected therewith, he shall be guilty of a misdemeanour.

" LV.—That with respect to the offenders whose names or residences are not known, any officer, agent, or servant of the Company, or any constable, police officer, or servant of any railway company, along or near to whose railway any electric telegraph, or any of the apparatus thereof, or any part thereof respectively, shall or may be erected or placed, or any constable or police officer, and all persons called by any such officer, agent, servant, or constable, as aforesaid, to his assistance, shall or may seize or detain any person who shall or may have broken, injured, or obstructed the working of any electric telegraph of or belonging to the Company ; or any of the wires, standards, instruments, apparatus, or other parts of any such electric telegraph, or who shall have committed any other offence against the provisions of this Act, and whose name or residence shall be unknown to such officer, agent, servant, or constable, and shall or may convey such offender with all convenient speed before some justice, without any warrant or authority other than this Act ; and such justice shall proceed with all convenient speed to the hearing and determining of the complaint against such offender."

I may instance, that, in November last, the wires of this (the Magnetic) Company were cut during the riots at Wigan. Having obtained information, I caused three men, Peter Moorfield, Peter Fairhurst, and Thomas Bradshaw, to be apprehended, and obtained their committal to gaol to await trial.

We proved the charge against them at the Birkdale Sessions, in January, 1854, and a verdict sentenced them to six months' imprisonment, with hard labour.

Answer 50*th.*—RECEPTION OF BUSINESS FROM THE PUBLIC.—All messages are handed to the Company, written upon printed forms provided for the purpose. To all good customers small books of forms are issued. Larger books lie at the places of general resort (such as the exchanges, reading-rooms, etc., etc.) ; and any casual customers find forms ready at the Company's offices upon counters of a height suited for writing, when standing, and subdivided into spaces, with fluted glass screens between each, to prevent any person seeing another's message.

The Company's cashier quickly counts the words in the body of the message (the address not being included, but passing free), endorses the message, and writes a receipt of the amount ; the customer is handed the *receipt*, upon the money being paid. Parties sending messages are advised to write them distinctly ; and the cashier reads the message, in order to see that the writing is legible, before handing it through to the instrument room.

The cashier enters upon a list, opposite to the consecutive number of the message, the amount received ; and, on being passed through to the instrument room, the lad receiving the message marks the number upon a similar list, and sends the message to the instrument for which it is intended. The clerk at the instrument then dispatches it to, or towards its destination, receiving an affirmative or negative signal after each word ; if the latter, the word is repeated, not having been rightly understood by the receiving clerk at the distant station. So commencing the message, the *sending* clerk signals the number of words the message contains (previously inscribed on the paper by the cashier), and, as soon as completed, the *receiving* clerk's writer counts the number of words received, to see that the message is correct as to length ; and, as will have been seen, the " understand " or " not understand " signals after each word, check the words themselves— admitting, when the system is carefully carried out, of little possibility of mistake.

In the foregoing I have embodied the routine observed in our *chief* stations. In small stations, where there is no great influx of messages, the checking is not carried out to such an extent.

APPENDIX

As soon as the message has been sent, it is returned to the checking lad, who files it, and draws his pen through its consecutive number, to intimate that, as far as the due forwarding is concerned, the Company have performed their duty, and it is his business to see that the signal clerk has endorsed upon the document the time at which he sent it, the station to which he signalled it, and his initials. By such an arrangement all chance of a message being mislaid is avoided; as, if the communication is not returned in a quarter of an hour, to have its number marked off the list, it is the duty of the checking clerk to enquire after it, and to ascertain why it has not been dispatched.

Very little time is lost in such an arrangement, and the chance of error of any nature greatly diminished.

CELERITY OF TELEGRAPHING.— Messages are forwarded upon a pair of wires by needle telegraph, at an average rate of $27\frac{1}{2}$ words, of five letters each, per minute ; and the more expert operators can pass 35 words in a minute ; or as fast as the writer can possibly take them down.

The time occupied in forwarding messages, on an average of a thousand, between Liverpool and London, comes out $4\frac{1}{2}$ to 5 minutes each (including any occasional delays that arise) counting from the time when handed to the Company by the customer to the completion of their reception in the London office.

I cannot give any average for delivery, as that depends entirely upon the distance from the office to the address. Messengers are always in waiting, and one is immediately sent off with the communication in a sealed envelope upon arrival.

The order of precedence is determined by priority of handing the message ready written over the Company's counter to the cashier.

Instances, well authenticated by the customers themselves, have occurred where the whole process from the handing of a message to the Company, to its delivery to the party to whom addressed, has been accomplished in a minute and a half.

But this can only happen when the receiver's office is close at hand, and an instrument entirely disengaged at the moment the message was handed to the Company.

Answer 51*st.* —OFFICIAL BLANKS FOR MESSAGES REQUIRED TO BE USED BY THE PUBLIC.—No. If messages are brought into our offices on plain paper, the person bringing such is requested to copy the communication upon the printed forms provided. Otherwise the message would be refused by the Company ; though this is a contingency that does not arise, owing to the requirement being made without exception by each company.

If the customer cannot write, one of the Company's clerks copies the message, reads it to the customer, keeps the original, and obtains

446

APPENDIX

the signature or mark of the person at the foot of the Company's paper. The message is then sent, the Company being freed from onus.

Answer 52nd.—Printed forms have been used from the establishment of the telegraph.

It has been found requisite for many reasons. In the first place, the difference of cost between plain and printed paper is of small consideration, as compared with the necessity that would devolve upon the Company to explain to customers the mode of arranging a message upon paper.

The defined position of the address from and to, and of the body of the message, materially aids the instrument clerk in forwarding the communication.

Moreover, for the security of the Company as a trading concern, we consider it necessary to embody certain conditions and stipulations upon which alone we receive messages.

We also caution customers against indistinct writing, and disclaim all responsibility for the same, and require, finally, the *signature* of the customer in authentication of his message, and as *subscribing to the Company's conditions.*

INSURANCE FOR CORRECTNESS OF MESSAGES.—In the stipulations subscribed to, we mention that in order to ensure accuracy, it is necessary that communications should be repeated from the station to which sent ; and that an extra charge of half a rate is made for such repetition, the Company holding themselves responsible to the extent of £5 upon such repeated messages. Further, another clause provides for the insurance of valuable messages to any amount upon payment of one per cent. upon the amount insured.

This latter regulation is *never acted upon* ; and that of repetition seldom. No error has occurred in repeated messages.

When parties, as is frequently the case, hand us messages in French, German, Indian, or other foreign languages, we take them at the risk of the sender as to accurate transmission of the words.

Code signals for various descriptions of messages, and cyphers for short words of frequent occurrence, such as " the," " from,'' " and,'' "to," " in," " on," " you," " yes," etc., and for terminations, such as " tion," " ing," " ment," are much used by the staff to facilitate transmission.

The Company's customers (especially the stock and other brokers, and banks) use cypher to a great extent, one message out of four, on an average, being so written, the regulation being that code words shall not exceed two syllables in length, in order to prevent an abuse of the system by the introduction of words of a nature or length likely to delay the Company's operators.

Of course the correspondents arrange their codes between themselves, each having a key, so that on arrival of the message at either end, it is translated.

Answer 57th.—WHEN COPIES OF MESSAGES ARE GIVEN.—We are very seldom called upon to furnish copies of messages. The Company in all instances refuses to do so, unless upon the application of the person to whom the message is addressed ; and in all cases requires a satisfactory reason.

MESSAGES RECEIVED AS EVIDENCE IN COURT.—We produce messages in courts of law upon an order of the sitting-judge or magistrate ; they are received as evidence in a court of law when recorded upon oath from the Company's servants connected with the reception of any communication so referred to.

FILING OF DUPLICATES OF MESSAGES.—We keep a duplicate of each message (taken by means of carbon paper, in manifold), each duplicate being numbered and put by with the others at the end of each day--the parcels being deposited in hampers sealed, and in a cupboard under lock and key. At the end of two years, the messages are damped and mashed, until all trace of writing is lost.

Answer 58th.—COST OF STATIONERY. — Our forms were much larger, and on a finer description of paper than at present used. But after a conversation with Mr. Shaffner, I have thought it best to change the size of the forms, and also the paper used ; obtaining the present stationery at a greatly reduced rate.

Cost of Forms as under :—

Forms on which messages are written by customers, printed on both sides, with particulars, regulations, etc.	Five shillings and three pence per thousand, or $1.26.
Received message forms upon which communications are written for delivery.	Same price as above.
Envelopes, printed with the words "Immediate," "By Magnetic Telegraph," and adhesive.	Five shillings per thousand, or $1.20.

Answer 59th.—Cannot give average—so much variation as to size, locality, etc.

Answer 60th.—TIME OF CLERK LABOUR.—Duty of clerks and messengers averages nine hours during day, and eight hours when on evening or night duty at the principal stations.

The older clerks take night duty in rotation. The younger clerks are always kept to day or evening work.

Answer 61st.—No.

Answer 62nd.—RESPONSIBILITY FOR ERRORS IN LAW.—We have never lost damages in any court—the question of responsibility has been occasionally tried. The most recent case is one in which the British Telegraph Company were implicated — the plaintiff being beaten.

I subjoin a copy of a press report from the *Glasgow Chronicle,*

APPENDIX

dated June 28th, 1854, of a suit instituted by Messrs. Dick & Martin, shawl manufacturers, Paisley, against the British Telegraph Company. Subjoined verbatim :—

"A manufacturing firm in Paisley, having lately sent a message by telegraph to their London agent, 'return all printed squares above five shillings,' and the message having been delivered with the word 'scarfs' substituted for 'squares,' which led to two days' delay in the squares arriving in Paisley, the manufacturers brought a small debt action against the telegraph company, and their agent, Mr. William M'Intyre, jun., for £12 damages, as loss sustained by them in consequence.

"The defendants denied liability, pleading the message order as a special contract, which contained a condition, declaring that they would not be liable for mistakes. The sheriff took the case to avisandum for a week, and on Thursday last pronounced his decision, sustaining the defence."

The *Chronicle* remarks, that the decision must be regarded as of great importance, considering the continual risk of small mistakes occurring in the transmission of messages over long distances with several "repeats," and the ruinous amount of damages in which telegraph companies, if liable, might be involved. The agent for pursuers in this case was Mr. John Guy; for defendants, Mr. Thomas Campbell.

In this instance, the message had to travel over several companies' lines. The British Company's wire not extending beyond Manchester.

Legal advisers, as a rule, do not like bringing forward such cases in the face of the conditions suscribed to by their clients upon the Company's paper.

I consider that a claim cannot be well established except in a case of a repeated or insured message, as a claimant would be met with "Why did you not insure, if your message was of such consequence, seeing that the Company has a special provision for such cases?" And in the case of any mistake or non-delivery of an insured message, the Company would be willing to pay, upon proof, without going to law.

Answer 63rd.—FREE MESSAGES.—No free messages, except those purely on the business of the Company, and sent by the officers of the Company, are allowed.

When directors or shareholders wish to send messages, they pay the ordinary tariff like any other customer, handing the message over the counter in the usual form.

Answer 64th.—TARIFF FOR PRESS NEWS.—Our rates to the press are far less than to the public. There are two distinct systems in operation as regards the transmission of intelligence for the news

APPENDIX

rooms and papers. Either they can supply themselves from their own agents, or contract for a supply from the Company. In the former case, a universal tariff of sixpence per line of nine words is charged, whether between London and Liverpool, London and Glasgow, or London and Cork, etc., and by this system, of course, *special* news can be forwarded.

In the latter, the Company having established agents and reporters in the various towns with which we communicate, the press are furnished daily with a column and a half to two columns of the latest political, foreign, commercial, and market news, in return for a payment amounting on the average to a little over a farthing a line. But, of course, in this case the news forwarded is perhaps distributed in duplicate among the three or four news rooms and dozen papers of the town.

Answer 65th.—REGULARITY OF BUSINESS.—This Company, since the commencement of *working*, in February, 1852, lost four messages. In each case the amount paid was returned ; none were insured. In one case an action was brought, but could not be sustained.

Answer 66th.—If, under any circumstances, a message is not sent in half an hour after being handed to the Company, the superintendents are instructed to send word to the customers, giving the reason why the message has been delayed ; it is then at the option of the party to withdraw his message and receive back the amount paid, or to allow the Company further time to transmit it.

Answer 67th.—As previously mentioned, Government communications take precedence.

Private (paid) messages come next in order, constituting the bulk of the Company's business. Next to these rank news messages, whether forwarded by the agents of the Company, or of the press. In any great emergency, however, where a prompt order is necessary, a business message (usually *last* in order of precedence), is sent by an officer of the Company before anything else, with a peculiar code or cypher prefixed to denote the extreme urgency of the message ; and if any communication should be sent with such prefix without necessity, the officer sending it would be subjected to dismissal.

I may mention that the Company's news messages are usually passed at hours in the day when the commercial business is slack.

Answer 68th.—PRE-PAYMENT REQUIRED ON MESSAGES.—Prepayment is required upon all messages forwarded by the public ; but good customers—sending many messages in the course of a day—are not usually called upon to pay at the time of sending each message ; but an account is made up in the afternoon and sent round by a clerk, whose duty it is to collect in money for the cashier.

In cases of great exigency or emergency, persons are sometimes allowed to send messages "to be paid for by the receiver" ; and

APPENDIX

answers are frequently pre-paid by those sending the messages re-quiring answers.

Answer 69th.—No. Answered in 63.

Answer 70th.—OPERATING DEPARTMENT.—Three clerks to every two instruments, *i.e.* an instrument while *sending* requires *one* clerk ; and when *receiving*, two, one to read off, and the other to write down from dictation.

Answer 71st.—The apparatus belonging to, and used by, this Company, is the magnetic telegraph (the practical application of magneto-electricity to telegraphic apparatus), and of course consists in principle of a needle telegraph, worked by the inductive influence exercised by magnets upon electro-magnetic coils, when placed in propinquity to the poles of the permanent magnets.

The electro-magnetic coils are so arranged that their cores of soft iron serve as a keeper (*in facto*) to the permanent magnet.

Professor Faraday is accounted the discoverer of the *principle* ; and after many unsuccessful attempts to apply it to practical purposes, Steinheil, of Munich (I believe), was able to construct a magneto-electric machine to work between Munich and Bogenhausen.

Subsequently Professor Wheatstone endeavoured to apply it, but failed, except so far as relates to ringing bells by an apparatus familiarly known as a "thunder-pump," from the pumping action of the machine, which consisted of a lever-handle, at the end of which two large electro-magnetic coils were fixed—the axis allowing the coils to be attracted against a magnet ; this apparatus engendered a strong current of electricity for ringing bells, but no indicating apparatus was contrived.

Mr. Henley afterwards managed—by arranging a magnet-needle between a pair of horns of soft iron, projecting from the poles of a pair of small electro-magnetic coils—to obtain with the motion of the lever a backward and forward movement of the needles.

Since that time, various improvements have been introduced by my brother Charles and myself. The apparatus is found to answer every purpose as far as speed, certainty, invariability of action, and small cost of maintenance is concerned. The improved apparatus is, moreover, the only telegraph that I am acquainted with that works satisfactorily in connection with underground wires, as in its present form it is not at all affected by the induced currents that affect other apparatus. Cost varies from £15 for small to £36 for large.

Answer 72nd.—Mistakes are of very unfrequent occurrence with us —averaging about 1 in 2,400 messages sent by this Company, and two out of three are occasioned by the indistinct writing of customers. Others are caused by the similarity of sound between certain words, such as "hour," "our," —"one," "done," etc. The immediate move-ment of our needles and their *dead-beat* (*i.e.* the absence of all vibration and oscillation) greatly tend to prevent mistakes. In the ordinary galvanic telegraphs of Cooke, Highton, and others, the needle sways to and fro after each beat, occasioning confusion between letters, which are formed by combinations of *"beats."* We also employ clerks in charge to check over all messages received or transmitted, to see that the context is correctly rendered, and the words rightly spelt.

Answer 73rd.—At first we repeated all paid messages, but I found it led to more frequent error, as the clerks relied so much upon the repetition to correct any error that they signalled in the first instance carelessly ; and in any pressure of business the repetition being made still more hastily, they more frequently committed blunders.

I do not quite understand the latter part of the question, but suppose that Answers 56-7, 66-7-8-9, will also apply to this.

Answer 74th.—As mentioned in Answer 50, each word, on being forwarded through the instrument, is succeeded by an affirmative or negative signal from the receiving station. If the former, the word next in order is sent ; if the latter, the word just sent is repeated.

Answer 75th.—The plan of insurance has been prescribed by this Company for two years and a half; but, as mentioned in 52, the principle has not been adopted by customers.

Answer 76th.—If called upon to insure, the Company would refuse to take such responsibility beyond their own lines, having no mutual arrangement with other companies on the subject. And, in any case, it would be highly dangerous for one company to undertake responsibility for the working of another company, especially if the slightest ill-feeling existed on the part of either, as one might take advantage of such a system of insurance to ruin the other.

Answer 83rd.—PRESERVATION OF WIRE CABLES.—In reference to this question, I may observe that tar is found to act as a great preservative in connection with iron, when immersed in salt water ; hence, in making a cable, it is advisable to saturate the hemp surrounding the conducting wires with Stockholm tar, prior to laying on the outer casing of protecting wires. With a cable so constructed, it is advisable to select as *sandy* a bottom as possible ; for the cable we laid between Donaghadee and Portpatrick, made on this plan, is found to have surrounded itself, when passing through sand, with a hard concrete, impervious to moisture, composed of sand and the tar that had oozed from the cable. There is an objection to galvanized wire, owing to its brittleness, interfering with the process of manufacture, and with the laying, by continual snapping ; the galvanizing appears to make the wire in parts assume the crystalline condition in lieu of the fibrous.

Answer 91st.—CABLES NOT LIABLE TO INJURY BY LIGHTNING. —A submarine cable, in connection with the underground system, is

APPENDIX

not, I apprehend, liable to injury by lightning, as such a cable would offer no more inducement, but at the same time far greater resistance to the exit of electricity through the insulating medium, from its wires, than the underground wires would evince. In the first place, the gutta percha coating of the submarine wires is considerably thicker than that around the land wires; moreover, each wire in the cable is surrounded in addition by a thick non-conducting layer of closely-packed tarred yarn.

The electricity, if ever it managed to get into the underground wires in any dangerous quantity, would naturally seek an equalization by the easiest path, and would make its way through the insulating coating of the subterranean wires to the moisture surrounding them, in preference to the cable.

It would, however, be extremely difficult for the electricity to pass into the wires, which are never brought above ground, save in the offices, when in connection with an instrument; and lightning would be opposed on entry by the thin coil-wire of the apparatus.

If submarine cables are used in connection with *overground* wires, it would be advisable to use every means of arresting, or rendering nugatory, a discharge of atmospheric electricity.

I should think several short coils of thin wire, consecutively interposed in the circuit on either side, would produce the required effect, especially when placed in an *earth* box; for any electricity that could pass without facing the coils and escaping to earth *in transitu*, would be comparatively harmless in the cable.

Answer 94th.—ELECTRIC CIRCUITS FOR SUCCESSFUL WORKING. —I observe, in my brother's answer to this question, that he has merely referred to our existing arrangements with regard to circuits, and I would therefore add a few remarks.

It has been found in England that none but large towns yield any profit upon the working expenses of a telegraph; we therefore determined, in the outset, not to extend our wires to any points where a profit could not be obtained; hence our station list is small, and does not require more than five or six, or, at the utmost, eight stations in a circuit.

It is not, I consider, advisable to have too many stations in a circuit, continual interruption arising, combined with difficulty in obtaining that prompt attention to calls requisite.

Moreover, from the many connections, a circuit with a number of stations at all approaching that mentioned, would be much more likely to get out of order from its integrity being disturbed.

In addition to this, an operator, in case the wires were occupied when he wished to forward a message, could not tell how long he might have to wait before the other stations would be sufficiently dis-

453

APPENDIX

engaged to allow him time to obtain the attention of the station he required, and to send his message ; and the *calls* become very complicated when more than eight stations are in a circuit.

Many towns in this country have no demand for speedy communication by telegraph, although the charge for messages is low—about a shilling for twenty words (address not being counted, and porterage within a mile of station free) per hundred miles. This want of a demand is chiefly due to the facilities afforded by the postal arrangements, which are speedy and excellent, and partly to the steadiness of business and non-existence in many towns of any speculative trade.

We possess arrangements of a simple nature, ready for use whenever needed by the Company, by which, even with a badly insulated line, any number of stations can be linked together, in immediate correspondence to and fro with one another, whether on the main line or branches.

Answer 98*th.*—GOVERNMENT LINES IN ENGLAND.—The use of the telegraph in connection with the Government and the police might be much more *systematized*, but at present great desultory use is made of the various telegraphs in this country by Government, relating to the army, navy, and commissariat. In the southern districts Government has had several lines constructed between the headquarters of Government and the chief naval depôts and arsenals, and have their wires worked by their own staff.

ARREST OF FUGITIVES FROM JUSTICE.—In many instances we have aided the police in arresting fugitives from justice. Sometimes telegraphing the exact dress, height, and personal appearance, with any special marks for identification of the party sought after.

In each case the Company is merely the *carriers* of orders and instructions between the police of one city and another.

In the shipment of troops to the East, and prior to the departure of the naval squadrons, Government made continual use of our wires, orders, countermands and directions following one another.

Answer 99*th.*—BANKING BY TELEGRAPH.—The banks generally pass through certain codes known only to themselves and their correspondents —varying with each day. Such code-signals precede the *body* of the message. By this check on frauds, banks daily transact a large amount of business—chiefly relating to returning bills or stopping local notes. Sometimes remittances of £20,000 and even £30,000 ($140,000) are made by means of advice by telegraph from one bank to another. They frequently also inquire as to respectability of parties—managers holding applicants in conversation till the answer arrives.

MESSAGES RECEIVED AS EVIDENCE IN COURT.—Messages are received as evidence in court. The signature of the party sending

454

the message being sworn to, and also that of the party receiving (entered on the messenger's ticket).

There is no law compelling evidence on the part of a telegraph company ; but in all cases where necessary, that I am aware of, evidence has been given freely.

APPENDIX V

THE first paper on the subject of submarine telegraphs was read before the Institution of Civil Engineers, by Mr. F. R. Window, on January 13th, 1857. This was mainly a somewhat *ex-parte* record of events. Bright, on being called on by the chairman (Mr. I. K. Brunel, F.R.S.) to lead off the discussion, made the following observations [1] :—

Mr. Charles Bright remarked that some additions were necessary to render the paper a complete report of the progress of submarine telegraphy up to the present time. The proportionate strength and weight of the different cables, and their construction ; the depth and character of the bottom at the various sites, and the manner of paying out, as also some information as to the power used in the experiments recorded in the tables, should have been given. The proportions between the power of the current and the different lengths and kinds of wire should, likewise, have been examined.

The chief point suggested for discussion was the difficulty of working, at such a rate as should be commercially successful, through such a length of cable as that now being constructed to connect Europe with America. There were reasons to believe that the effects of the phenomena of induction and retardation were exaggerated. The electrical conditions of an underground wire coincided with those of a submarine wire. It was true that electric currents, employed in the usual way with the common needle and recording instruments, were so retarded, that if worked quickly the signals were confused or blended together ; but this could be effectually dealt with.

The first important English underground line coated with gutta percha was that laid by the Magnetic Company, in 1851, between Liverpool and Manchester. On connecting the wires at each end, so as to form a continuous length of above one hundred and forty miles, the effects of induction were clearly seen, and a strong secondary current issued from the wire on breaking contact. But Mr. Henley's

---

[1] *Minutes of Proceedings Inst. C.E.*, vol. xvi.

APPENDIX

instruments met the difficulty, and promised, with some modifications, to overcome any extension of it likely to arise in the longer lines then contemplated. The novelty in these instruments was the mode of generating the current, its proportions as regarded intensity and duration, and in the method of acting on the needle. In the old plan a galvanic current was sent in one direction only for each signal, the needle, or armature, being brought back to its place of rest by a weight or spring. This, independently of the inductive effect, was objectionable, as that weight, or tension, had first to be surmounted by the current, and thus the power available to produce a signal was only the difference between the force of the current and the retaining power of the weight or spring. In the system alluded to the current was derived from a permanent magnet, and was (proportionately) more intense and rapid in its effect. The needle was equally poised on all sides, and worked as well horizontally as vertically. For each signal the needle was first deflected by a current in one direction, and then reversed by a current in the opposite direction, the currents being of precisely equal force and duration. This equality of time and force was a very important feature, for any difference was fatal to working through great lengths of wire affected by induction. In an instrument since contrived to work with reversed currents from a galvanic battery, in which the duration of the currents varied according as the signals, or pauses, were longer or shorter, a short current was found not to be able to surmount the residual effects of a long one, and thus a proper neutralization did not take place. The increase of induction, by a current of great duration, was self-evident, and might be practically seen by charging a well-insulated wire for some time ; the return current, or discharge, had then more power than a brief current from the battery itself on short circuit.

On the completion of an underground line from London to Manchester, no cause for discouragement was found, as, with some alterations, the instruments worked admirably. Since then the underground system had been extended, without any inconvenience as regarded electrical working. Indeed, instead of being limited, the longest circuits had been worked at the ordinary business rate—about twenty words a minute. Now each letter, with the reversing system, required an average of four currents to be sent, and taking five letters to a word, there would be four hundred currents per minute in practice, which would, from their length, only allow of ten currents being sent in the same time, if the theory advanced was sound.

In considering the mode of working through still greater lengths, it was obvious that the principle would require a different apparatus. Partly in conjunction with Mr. Whitehouse, whose researches and ingenious instruments were so well known, he engaged in a series of experiments, which had produced most satisfactory results. Some of

457

APPENDIX

these experiments were repeated a few months back, for the satisfaction of Professor Morse, through two thousand miles of underground wire, connected so as to form a continuous circuit, terminating at both ends in the earth. Intermediate instruments were placed at each loop to test the action of the electrical waves throughout the entire length, without any creeping of the current from wire to wire, and galvanometers were inserted at the distant ends. Signals were then clearly defined at the rate of ten to twelve words per minute. Two large induction coils,[1] three feet in length, excited by a powerful "Grove" battery of fifty pint cells, but connected for quantity in sets of ten, were used to generate the currents. The power of working through very great lengths depended upon the force, duration and proportions of the current used. From all that had been shown in the paper, he contended that no difficulty was likely to arise that could not be effectually dealt with in working from Ireland to Newfoundland by the proposed Atlantic cable.

Subsequently the discussion was adjourned, and in reply to some of the speeches that followed on the next occasion, Charles Bright made these observations :—

Mr. Bright remarked that there was no such thing as two thousand miles of telegraph wire in one continuous length. In the experiment he had alluded to, ten separate lengths of underground wire, between London and Manchester, were connected together. As it was thought that it might be inferred that the passage of the current was assisted by lateral induction, Professor Morse had tested this point very carefully, and had found that the inductive effect from wire to wire was so slight that a delicate galvanometer was not at all affected by it. It might, therefore, be fairly assumed that no induction from wire to wire would affect the working of the instruments. He might mention that, partly in conjunction with Mr. Whitehouse, he had tried some experiments on seven hundred miles of underground and submarine wires, in one direct length, from London, round by Scotland, to Dublin. This was the greatest length yet tried in a direct line, working to the earth, and in direct circuit with the wire—and for that distance the results were most satisfactory. The coils used were about twelve inches in length. The coils employed in the experiment with the two thousand miles of connected wire were thirty-six inches

---

[1] The Induction Coil, stated by the Rev. Professor Callan to have been discovered by him in 1836, led to his devising, in 1837, an instrument for getting a rapid succession of electrical currents from the coil, and thus completing the coil, as a machine, by which a regular supply of electricity might be furnished.—EDITOR.

APPENDIX

in length, and were excited by a more powerful battery. Now, if they could work through seven hundred miles of direct wire, with the earth intervening, and could send an electric sensation through two thousand miles of connected wire with four instruments in London, and four galvanometers in Manchester, he could not see what there was to prevent the working, successfully, through a direct line of two thousand miles. Professor Thomson, of Glasgow, was at one period a great opponent of the theory of the practicability of working through long distances, but he had seen reasons to change his opinion on the subject, and was now one of its warmest supporters. The instrument which had been shown, so arranged as to send alternate galvanic currents by the motion of a finger-key, was defective in principle. There was this disadvantage in it—that instead of each current being of equal length, the first current to deflect the needle was of one length, and the second to bring back the needle was not always of the same length, so that the sensation through the wire was not perfect. For this reason, when working at the ordinary rate of transmission through a long submarine wire, the pause between a short signal and a long one would not be sufficient to enable the short one to overcome the residual effect of the long one ; when they charged a wire for a long signal, for, say, $\frac{1}{10}$th or a $\frac{1}{4}$ of a second, a shorter signal would not be able to overcome it. It was a matter of duration, and of strength of the current.

The answer of the author with respect to the three hundred and sixty cells was not satisfactory in an electric point of view. It did not appear whether the intensity of the current had been increased, as the length of wire was increased from four hundred miles to eight hundred miles, and from eight hundred miles to sixteen hundred miles, or whether the same three hundred and sixty cells had been used in each case. Should the latter view be correct, as he imagined, then a different result could not have been expected, for the current would necessarily be slow.

The tables from which arguments had been extracted against the successful working of long lines of submarine wire showed forcibly the weakness of those arguments. It has been stated that Professor Wheatstone proved the passage of an electric current to reach the enormous speed of two hundred and eighty-eight thousand miles per second, and the table gradually descended to only eighteen thousand miles with other experiments on the same class of conductors. If the nature of the exciting powers had been recorded, the statement would have clearly demonstrated that different proportions of power produced very different results. The first experiments were made with high-tension electricity, the last with currents of greater quantity, and hence the difference. Reasoning from theory, without regard to practical results, taught that a very much greater rate than that

APPENDIX

named in the paper would be attained in the line from Ireland to Newfoundland.

Later in the evening a number of points were raised as regards the programme for the Atlantic cable enterprise, to which Charles Bright was looked to for information. Regarding the main point he spoke as follows :—

Mr. Bright, in reply to questions from different members, said that the Atlantic cable was then in process of manufacture, partly at the works of Messrs. Glass, Elliot & Co., at East Greenwich, and partly at Messrs. Newall's works, at Birkenhead.

APPENDIX VI

On the return of the Atlantic Cable Expedition of 1857, the *Morning Post* gave expression to the situation with a brief *résumé* of the events, in the course of the following leading article :—

So far as the narrative of a severe disappointment can be satisfactory, the public will read with satisfaction the report published in our columns yesterday respecting the accident to the Atlantic Telegraph cable. The hopes we had entertained of linking the Old and New World together by a chain of telegraphic communication have indeed been frustrated. Elaborate and costly preparations are to some extent wasted, and the completion of the work is postponed, perhaps for an entire year. On the other hand, much valuable experience has been gained ; and the misfortune shown to have arisen solely not only from a complete accident, but one against whose recurrence it is easy to provide. The worst that can be said of the casualty is, that it is provoking, as causing delay and adding to expense. With regard to the ultimate success of the undertaking, the occurrence is in no way discouraging, but rather the reverse.

From Mr. Bright's well written account, we learn that every description of obstacle to which the enterprise was liable, had been met and overcome. The cable, as he states, " has been laid at the expected rate in the great depths, its electrical working through the entire length has been most satisfactorily accomplished, while the portion laid actually improved in efficiency by being submerged, from the low temperature of the water and the close compression of the texture of the gutta percha." The process, moreover, was carried on with little interruption from the condition of wind or wave. Though the sea was heavy, almost from the outset, it was found practicable to continue paying out the cable ; and though the accident which stopped proceedings arose primarily from the state of the weather, it was immediately occasioned by a piece of clumsiness in the manipulation, which will be carefully avoided when the undertaking is resumed. Success is now shown not to depend upon the rare fortune of finding a fortnight's calm weather on the Atlantic contemporaneously with the execution of the work. Any moderate gale will produce no serious impediment, provided the management of every part of the operation is confided to assistants possessing an adequate amount of mechanical skill.

APPENDIX

Before returning from the frustrated expedition an opportunity was taken of proving by trial the possibility of accomplishing in mid-ocean the delicate operation of joining the two ends of the divided cable. When the line was half laid down, as our readers know, the coils on board the *Niagara* would be exhausted, and the remainder furnished from her consort, the *Agamemnon*. The problem whether a safe junction could be effected between the end of one section and the beginning of the other while paying out in deep water was one of no small importance. The preparations made for that purpose were considered ample. But in a case like this nothing is so satisfactory as an actual trial, and this was performed with complete success. The two ends on board the *Niagara* and *Agamemnon* respectively were joined together, and the splice let down to the bottom into soundings of 2,000 fathoms during a heavy sea. Some miles of the spliced cable were afterwards payed out from each vessel without difficulty.

The directors of the Company, after fully investigating the occurrence, have issued a report, in which they hold out sanguine expectations of future success. "Sufficient information," they say, "has already been obtained to show clearly that the present check to the progress of the work, however mortifying, has been purely the result of an accident, and is in no way due to any obstacle in the form of the cable, nor of any natural difficulty, nor of any experience that will in the future affect in the slightest degree the entire success of the enterprise. The only sudden declivity of any serious magnitude (from 410 fathoms to 1,700 fathoms) had been safely overcome, the beautiful flexibility of the cable having rendered it capable of adapting itself, without strain, to circumstances which would probably have been its ruin had it been more rigidly constructed. The combined influence of the low temperature of the water, and the compression of the pores of the insulating medium, had practically shown that the action of a telegraphic cable, so far from being impaired, is materially improved by being sunk in deep water. These and all other circumstances which have been brought out by the recent expedition have made more and more cheering and certain the prospects of complete success on the next occasion."

This sanguine view of the case is fully justified by the incidents that have occurred, and will, we have little doubt, be borne out by the event.

APPENDIX VII

SUBSEQUENTLY to the Atlantic Cable Expedition of 1857, and prior to starting out again in 1858, Charles Bright sent in a report to the directors of the Atlantic Telegraph Company. This report comprises a full record of the events up to date, together with a statement of what was being done, besides the programme for future proceedings, and reads as follows :—

Early in the month of April, 1857, H.M.S. *Agamemnon* was placed at my disposal as your engineer ; and the fittings necessary to adapt her to the reception of the cable having been carried out with the utmost rapidity, she was moored at her station at Greenwich to take in the eastern half of the cable.

On the 14th of May, the U.S. frigate *Niagara* arrived in the Thames ; but, on calculating the space available for our requirements, it was found that considerable alterations would be necessary to suit her interior to our purpose. These were put in hand at Portsmouth, and she finally proceeded to Birkenhead, to receive her portion of the cable.

In the *Agamemnon*, by clearing her hold of the tanks and magazines, the available space allowed of the cable being made into one great coil, forty-eight feet in diameter and twelve feet high. In the *Niagara*, it had to be disposed in five coils, three in the hold, orlop-deck and berth-deck forward, and two on the berth and main decks aft.

The machinery for regulating the egress of the cable from the paying-out vessels was constructed with regard to the great depths of water to be passed over, the constant strain, and the number of days during which the operation must be unceasingly in progress.

The cable was passed over and under a series of sheaves, having the bearings of their axles fixed to a framework, composed of cast-iron girders bolted down to the ships' beams.

The sheaves were geared to each other, and to a pinion fixed to a central shaft, revolving at a rate three times faster than that of the sheaves ; two friction drums upon this shaft regulated the speed of paying out, and the grooves of the sheaves (which were fixed to their axles outside the framework and bearings) were fitted to the semi-circumference of the cable, so as to grasp it firmly, without any pressure by which it could be injured.

I need not here enter into the arrangements for splicing, buoying, guard-ropes, staff, lights, and other minor details of the expedition, nor into the causes which led to your resolution, that the laying of the cable should commence from Ireland, instead of from the centre, as was at first contemplated. On the 29th of July, the two ships, with the whole of the cable on board, met at Queenstown. On the 3rd of August, after uniting the two lengths, to test the conductivity of the entire line, and taking in coals and sundry stores, we started for Valentia, in company with H.M.S. *Leopard* and the U.S. frigate *Susquehanna*, two powerful paddle-wheel steamers, appointed to render assistance in case of need.

At Valentia, we were met by H.M.S. *Cyclops*, and on the 5th the end of the cable was landed at Ballycarbery strand from the *Niagara*, which lay in the bay about two miles distant.

An accident to the heavy shore-end cable shortly after weighing anchor on the 6th, deferred our final departure until the 7th of August·

For three days everything proceeded as satisfactorily as could be wished ; the paying-out machinery worked perfectly in shallow, as well as in the deepest water, and in rapid transition from one to the other ; while the excellent adaptation of the cable in weight and proportions to the purpose was most forcibly demonstrated by the day's work previous to the mishap, during which 118 miles of the cable were laid, for 111 miles run by the ship.

The details of the voyage from the 7th until the morning of the 11th, are fully set forth in the following extract from a report[1] made by me to the Board shortly afterward :—

By noon, on the 8th, we had paid out 40 miles of cable, including the heavy shore-end, our exact position at this time being in lat. 51° 59' 36" N., long. 11° 19' 15" W., and the depth of water, according to the soundings taken by the *Cyclops*, whose course we nearly followed, 90 fathoms.

Up to 4 p.m. on that day the egress of the cable had been sufficiently retarded by the power necessary to keep the machinery in motion at a rate a little faster than the speed of the ship ; but as the water deepened it was necessary to place some further restraint upon it by applying pressure to the friction drums in connection with the paying out sheaves ; and this was gradually and cautiously increased from time to time, as the speed of the cable compared with that of the vessel, and the depth of the soundings, showed to be requisite.

By midnight 85 miles had been safely laid, the depth of water being then a little more than 200 fathoms.

---

[1] Given in its place on p. 190.

APPENDIX

At eight o'clock on the morning of the 9th we had finished the deck coil in the after part of the ship, having paid out 120 miles; the change to the coil between decks forward was safely made.

By noon we had laid 136 miles of cable, the *Niagara* having reached lat. 52° 11' 40" N, long. 13° 1' 20" W., and the depth of water having increased to 410 fathoms.

In the evening the speed of the vessel was raised to five knots per hour; I had previously kept down the rate at from three to four knots for the small cable, and two for the heavy end next the shore, wishing to get the machinery well at work prior to attending the speed which I had anticipated making.

By midnight 189 miles of cable had been laid. At four o'clock in the morning of the 10th the depth of water began to increase rapidly, from 550 fathoms to 1,750, in a distance of eight miles. Up to this time 7 cwt. strain sufficed to keep the rate of the cable near enough to that of the ship; but, as the water deepened, the proportionate speed of the cable advanced, and it was necessary to augment the pressure by degrees, until, in the depth of 1,700 fathoms, the indicator showed a strain of 15 cwt., while the cable and ship were running five and a half and five knots respectively. At noon, on the 10th, we had paid out 255·miles of cable, the vessel having made 214 miles from shore, being then in lat. 52° 27' 50" N., long. 16° 00' 15" W. At this time we experienced an increased swell, followed late in the day by a strong breeze.

From this period, having reached 2,000 fathoms of water, it was necessary to increase the strain to a ton, by which the cable was maintained in due proportion to that of the ship.

At six in the evening some difficulty arose through the cable getting out of the sheaves of the paying-out machine, owing to the tar and pitch hardening in the grooves, and a splice, of large dimensions, passing over them. This was rectified by fixing additional guards, and softening the tar with oil.

It was necessary to bring up the ship, holding the cable by stoppers until it was again properly disposed around the pulleys. Some importance is due to this event, as showing that it is possible to lay to in deep water without continuing to pay out the cable—a point upon which doubts have frequently been expressed. Shortly after this the speed of the cable gained considerably upon that of the ship, and up to nine o'clock, while the rate of the latter was about three knots by the log, the cable was running out from five and a half to five and three quarter knots per hour. The strain was then raised to 25 cwt., but the wind and sea increasing, and a current at the same time carrying the cable at an angle from the direct line of the ship's course, it was found sufficient to check the cable, which was at mid-

APPENDIX

night making two and a half knots above the speed of the ship, and sometimes imperilling the safe uncoiling in the hold.

The retarding force was therefore increased at two o'clock to an amount equivalent to 30 cwt., and then again, in consequence of the speed continuing to be more than it would have been prudent to permit, of 35 cwt.

By this the rate of the cable was brought to a little short of five knots, at which it continued steadily until 3.45, when it parted ; the length paid out at that time being 380 statute miles.

I had up to this time attended personally to the regulation of the brakes ; but finding that all was going on well, and it being necessary that I should be temporarily away from the machine to ascertain the rate of the ship, and to see how the cable was coming out of the hold, and also to visit the electrician's room, the machine was for the moment left in charge of a mechanic, who had been engaged from the first in its construction and fitting, and was acquainted with its operation. I was proceeding toward the fore part of the ship when I heard the machine stop. I immediately called out to ease the brake and reverse the engine of the ship ; but when I reached the spot the cable was broken.

On examining the machine, which was otherwise in perfect order, I found that the brakes had not been released, and to this, or to the hand-wheel of the brake being turned the wrong way, may be attributed the stoppage, and the consequent fracture of the cable ; when the rate of the wheels grew slower, as the ship dropped her stern in the swell, the brake should have been eased. This had been done regularly before, whenever an unusually sudden descent of the ship temporarily withdrew the pressure from the cable in the sea.

After the accident the commanders of the vessels proceeded to Devonport at my request, the dockyard at Keyham affording many facilities for unshipping the cable.

At a subsequent discussion the prudence of making a second attempt in October was considered, but the difficulty of obtaining sufficient additional line, and the uncertainty of the weather so late in the year, were cogent reasons against the adoption of such a course. It was, therefore, decided to store the cable until next summer, and (having been granted the use of a vacant space of ground by the Government) four large roofed tanks were constructed to receive it.

The cable, which is in good condition, was discharged from the *Niagara* first, and has subsequently been unshipped from the *Agamemnon*. It has been passed through a mixture of tar, pitch, linseed oil, and bees' wax, in such consistency and quantity as effectually to guard against rust.

The buoys, chains, hawsers, and other stores and tools, are safely warehoused in the adjacent building.

APPENDIX

Immediately upon the return of the expedition, steps were taken to recover such part of the cable laid from Valentia as could be raised so soon as the equinoctial gales might be over.

The *Monarch*, a steamer employed upon the submarine lines laid between Orfordness and the Hague, and fitted with the necessary appliances for picking up cables, was at first understood to be at our service for this work ; but some delay to our plans for recovery arose from the fact that, at the time she was expected to be available, she was dispatched by the company to whom she belongs upon another duty, and it thus became necessary for us to procure and equip another vessel.

In the middle of October I proceeded to Valentia with the *Leipzig*, a paddle-wheel steamer of a sufficient capacity. After some hindrance by the gales which prevailed at that time, fifty-three miles of the small cable and four miles of the heavy cable were got up ; the remainder of the shore end was under-run, and is buoyed ready for splicing next year.

The sea and swell on that coast at this season are so unsuited to the work that the attempt to regain the remainder must be deferred for some weeks ; but if the contract which has been accepted by you is successfully carried out, it will be more satisfactory as regards risk of outlay than for us to renew the operation.

The recovered cable, which is in good order and fit for use again, has been delivered into store at Keyham.

Referring to the proposal to order a further length of three hundred miles of cable, in addition to the four hundred miles now in course of construction by Messrs. Glass, Elliot & Co., I would observe that while I anticipate that the appliances suggested by experience will enable us to lay the cable this year with much less slack than is expected, I am convinced that more allowance should be made for contingencies in laying a line of such extraordinary length.

It is doubtless a circumstance much to be lamented in the past history of our undertaking, that the period within which it was intended to be completed did not permit of more time being given to experiment, design and ultimate manufacture. It is also to be regretted that no time was allowed for rehearsals of various plans of cable-laying in deep water, respecting which there had been no previous successful experience.

"The result has been that experiment and practice have been mixed together in one operation ; and hence, although all concerned actively in the undertaking are now fully alive to the means which will, in all human probability, secure success on the the next occasion, yet great expense has been incurred without an adequate return, which might have been avoided had the needful time for experiment been available."

APPENDIX VIII

ON February 16th, 1858, Mr. T. A. Longridge and Mr. C. H. Brooks read a Paper at the Institution of Civil Engineers, on "Submerging Telegraph Cables." The following week, Mr. F. C. Webb read one before the same body on "The Practical Operations connected with Paying Out and Repairing Submarine Telegraph Cables."

The discussion on the two Papers was taken together. In the course of this, Bright made the following observations [1] :—

Mr. C. T. Bright thought that all persons who, like himself, were interested in the extension of telegraphic communication, must be deeply indebted to the authors of the Papers which had been read. The first Paper might be regarded as a survey of the question from a theoretical point of view, whilst the second Paper was a fit companion to the former one, from the practical information which it furnished.

Reverting to the former Paper, he would allude, in the first place, to the subject of resisters, which was an important point, relative to which frequent suggestions had been made. There was no doubt, that advantage would result from the employment of any appliance which would retard the egress of the cable, so as to allow of heavier and better-protected cables being laid in deep water. From the difficulties attending the use of buoys, the authors appeared to have discarded them from consideration, as being too bulky to be carried, but suggested resisters instead. He had himself been engaged in some experiments in this direction, and he had found that a plane surface 1 foot square offered a resistance, when drawn through the water at the rate of one mile per hour, of about $3\frac{1}{4}$ lbs. ; and with a surface of 4 feet, the retarding strain, at the same speed, would amount to nearly 56 lbs. Colonel Beaufoy's nautical experiments,[2]

---

[1] *Minutes of Proceedings Inst. C.E.*, vol. xvii.

[2] *Vide Nautical and Hydraulic Experiments, with numerous Scientific Miscellanies.* By Col. M. Beaufoy. 4to. London, 1834.

APPENDIX

upon the comparative resistance of solids moving through water, gave a similar result, and he had no doubt of the accuracy of that conclusion. At the rate of two miles an hour, the resistance experienced with a disc having an area of 1 square foot, equalled a pressure of a little more than 13 lbs. ; at three miles it amounted to 29 lbs. ; and at eight miles to 203¾ lbs. But there was a disadvantage in the use of resisters, in the form generally proposed ; especially in the case of the sudden pulling up of the ship from any stoppage in-board, from the occurrence of a kink, such as was met with in paying out one of the cables most recently laid. Should the egress of the cable be suddenly stopped, from this, or any other cause, it must be subjected to a great strain, until the vessel could be brought up, or else it would part. The pitching of the vessel in a heavy sea was also an important feature, in considering the advisability of employing resistance floats ; a remedy for these objections might be found, in so attaching them that they would be released upon receiving pressure in the direction that would be prejudicial, but it would be far better to do without them altogether.

Another point suggested in the first Paper was, the possibility of catching the end of the cable, should it break near the paying-out vessel ; relative to which the proposal had been made, that auxiliary vessels should follow the paying-out steamer, with the cable passing through a ring. He did not think that plan would be successful, but, on the contrary, he believed it would add to the chances of losing the cable.

Another question was that of compensating for the rise and fall of the ship. The authors appeared to hold the opinion, that it was of comparatively minor importance, because it would merely alter the form of the curve in a small degree, and only "increase the abscissa of the catenary, by the amount due to half the height of the wave." With a fragile and light cable, he considered it of more consequence than had been allowed, to compensate for the pitching of the ship ; but he thought it was far from necessary, or even desirable, under all circumstances, especially where it was possible to use light paying-out machinery ; and there was a difficulty in carrying it out, so as to meet every requirement, without adding materially to the weight to be set in motion.

The chief subject suggested in the first Paper was the comparative advantages of light and heavy cables, and this, he apprehended, would form the leading feature of the present discussion. He fully subscribed to all that the authors had argued, with reference to the importance of light cables for deep waters. The lighter, the stronger, and the cheaper they could be constructed, the better ; but he had not yet seen what might be termed a very light cable that fulfilled all the conditions required of it. In the Atlantic cable, a considerable step had been taken in that direction, as far indeed as it was prudent to go in the present state of experience. He thought a great deal more had been done, in the matter of light cables, than was known to the authors. The first experimental cable, between Dover and Calais, was composed of a copper-wire, coated with gutta percha; it was as strong, in proportion to its weight, and the depth in which it had to be submerged, as the cable alluded to in the Paper was to such depths as had to be traversed in the present day. Then there was the cable attempted to be laid between Portpatrick and Donaghadee, in 1852, by a company which had since been wound up. The insulating substance was gutta percha with india rubber, covered with common rope. Then there was another—properly to be called a light cable, although it had iron wires laid spirally outside the core—between Holyhead and Howth, which weighed about a ton to the mile; also, the cable laid between Varna and Balaklava, in 1855. Thus, a good deal had been done, practically, in light cables; but the results had in all cases been most unfortunate. The Dover and Calais experimental cable, in 1851, failed after raising the hopes of all concerned in it. The hempen cable in the Irish Channel never reached the shore. The light cable from Holyhead to Howth broke down shortly after signals were sent through it. And the Government wire in the Black Sea, although it remained in good working order during the period of the war, broke down when peace was proclaimed, as if aware that its mission was over. The authors recommended another form of cable altogether, and, to some extent, censured those who had had to do with the Atlantic cable for not having adopted it, because it had been invented three years before. Mr. Bright need not advert to the many experiments by which the form of the Atlantic cable was determined; but the kind which had been advocated in the Paper was considered among others, and rejected, as it always must be by any one practically experienced in the working of telegraphic wires. The authors prefaced their remarks upon cables with the observation that the conducting and insulating powers of cables did not come within the scope of their arguments. But, it seemed to him, that this embraced the whole gist of the matter. The cable which they so strongly supported was composed of a strand of iron wire, covered with its insulating substance, combining the conductor and the strength in one. That seemed, at first sight, to present many advantages; but when the conductor and the insulator were examined, which were, in fact, the very first considerations in constructing a cable, it would be found that the principle adopted was wholly defective, the conductor being altogether unfit for the purpose. The conditions, with regard to overground wires and underground and submarine wires, were very different. The possibility of working through an overground circuit, whatever the loss by defective insula-

APPENDIX

tion, was limited only by the resistance of the conducting power of the wire. In submarine and underground wires, on the other hand, the resistance was increased enormously by the induction of the surface of the metal and the surface of the gutta percha. He regretted the absence of Professor Faraday, who had explained the phenomena attendant upon long gutta-percha-covered wires so clearly at a former meeting. The great object was to get a good conductor. To employ an iron conductor, such as was recommended, would be to increase to an enormous degree the difficulties experienced in lines of this kind. According to Becquerel, the conducting power of copper was 100, of gold 93, of silver 73, and of iron only 15, so that iron was possessed of six times less conducting power than copper. To produce an equal effect, it would, therefore, be necessary to enlarge the dimensions of the conductor ; whilst the difficulty of obtaining perfect insulation being then greater, by having so much more surface to protect, the chances of leakage would be proportionately increased.

Before concluding his remarks upon the first Paper, he would allude to an intimation, partially thrown out also in the second Paper, that sufficient publicity was not given to operations of this kind. He could only say that no ground of complaint could exist on that score with regard to the Atlantic cable, inasmuch as the details of the manufacture of every portion of that cable had been published as fully as possible ; and as far as he was personally concerned, he had at times most freely given all the information it was in his power to afford. It had been stated in the second Paper, that the Atlantic cable was not tested under water, from the fact that the wire used was not galvanized. Now, the core of that cable was regularly tested from the beginning, by Mr. Whitehouse, the laborious and careful nature of whose electrical experiments was well known. It was tested with a battery series of five hundred cells, and the most sensitive instruments that could be obtained from the instrument-makers of London and Berlin, who were most celebrated for delicate electrical apparatus. The wire, when tested under water, where it had remained for some days, was sent to the works upon drums, so protected that it could not possibly receive any injury, and was immediately covered with a serving of hemp, prior to receiving the outer strands. The table giving the breaking strains of common and galvanized iron wire, would probably lead to further discussion. He had always understood that galvanizing the smaller sizes of wire injured them, and he imagined that the table must apply to the larger sizes, although it was not stated whether the circumferences given applied to bar iron or to made rope. The diary of Professor Morse had been alluded to, and it appeared the assumption put forward in the second Paper, that the cable was not properly tested,

APPENDIX

was grounded on it. It was true that, as they got further from the shore, in the Atlantic undertaking, the signals were weaker ; but as Professor Morse was not concerned in the practical part of the undertaking, he was not acquainted with the conditions, or the reasons for what was done ; indeed he was confined to his berth nearly all the time. A very low battery-power was purposely employed, for objects that would be evident to all acquainted with testing wires. It was desirable to continue signalling to the shore, at the same time that any falling off in continuity, or in insulation, should be discovered by the effect on the instruments ; as new lengths of cable were added from time to time to the circuit, there was of course some variation of signals, and the adjustment of the instruments was sometimes interfered with by the motion of the ship ; but there was nothing uncommon in the occurrence, and the supposition that a fault had passed unnoticed was altogether groundless.

APPENDIX IX

BRIGHT'S REPORT ON SUCCESSFUL TERMINATION OF 1858
ATLANTIC CABLE EXPEDITION

SOON after successfully completing the laying of the 1858 Atlantic cable, Charles Bright sent in his report of the proceedings of the immediately preceding expedition, which the Secretary of the Atlantic Telegraph Company forwarded to *The Times* a few days later, as follows :—

" To the Editor of the 'Times.'

" SIR,—Herewith I forward you a copy of the report of Mr. C. T. Bright, the engineer-in-chief of this Company, in reference to the proceedings during the paying out of the cable, which you may probably consider to be of interest to your readers.

" Yours truly,

" August 24th.   " GEORGE SAWARD, Secretary."

To the Directors of the Atlantic Telegraph Company.

GENTLEMEN,—On arriving at Valentia on the morning of the 5th inst., I forwarded to you by telegraph a brief report of the success which has attended the Company's endeavours to place Newfoundland in electrical communication with Ireland, and I have now the honour to lay before you fuller particulars of the operations carried out on board Her Majesty's steamer *Agamemnon*, which I have been unable to do sooner, owing to the pressure consequent upon the return of the expedition.

After our departure from Queenstown, at 2 a.m. on the 18th ult., we proceeded towards the rendezvous, which we reached on the night of the 28th, having been delayed by contrary winds and a head-swell. We found the *Niagara, Valorous* and *Gorgon,* which had left Queenstown on the 17th, waiting for us ; and on the morning of the 29th, the sea being smooth and the barometer standing at 30·15, the *Agamemnon* and *Niagara* were connected together by a hawser stern to stern; the end of the cable on board the latter ship was then brought by the boats of the *Valorous* to the *Agamemnon*, where the splice was finished

473

APPENDIX

by one o'clock, local time, our position then being lat. 52·8 N., long. 32·27 W., distant 938·3 statute, o 815 nautical miles from the White Strand Bay at Valentia.

Having veered out a sufficient length to bring the splice into the centre of the curve formed by the cable hanging between the ships, the hawser was released, and we proceeded in our course slowly, paying out slack freely for the first three hours, after which the speed of the ship was increased to four, and at 7 p.m. to five knots.

All went on well until 7.45 p.m., when, immediately after passing from the outside to the centre of the coil in the main-hold, the beginning of the first turn of the flake next below that in process of delivery was seen (on being exposed by the uncoiling of the cable above it) to be squezed between the side of the cone in the eye of the coil and the end of the piece of wood by which the leading in part of the coil was defended.

This injury occurred through the extent to which the coil was disturbed during the gales encountered in our previous voyage; although the whole of the upper part of the coil, which had been displaced to such an extent as to promise any difficulty in paying out, was removed and coiled on the upper deck abaft the foremast, it would appear that all the new cable which had been lately placed on the top of the main coil had shifted somewhat in the heavy weather, for it was necessary to rectify another defect arising from the same cause at a similar part of the coil soon after.

The old cable, which had been coiled for a longer time, and was more thickly coated with the mixture of tar and pitch, was not in the least degree disturbed.

When the defective piece had been passed under some of the turns of the flake (then paying out to the outside, in order to allow of more narrow examination than could be made in the centre of the coil where the cable was passing out of the hold), Professor Thomson reported that continuity had ceased.

On the cessation of signals I requested Captain Preedy to stop the ship, having placed Mr. Clifford to superintend the machine, so that as little cable might be paid out as was consistent with safety—Mr. Canning taking charge of the reinstatement of the injury, while Mr. Hoar attended to the dynamometer.

It is in great measure owing to the care of these gentlemen that no ill resulted from this critical mischance.

At 9.15 the fault was repaired, and shortly afterwards signals were again reported from the *Niagara*. We had at this time paid out forty-six nautical miles of cable from the *Agamemnon*.

The depth of water at the time of this stoppage was 2,030 fathoms, according to the nearest sounding.

By noon on the 30th we had paid out 135·8 nautical miles, being

APPENDIX

then in lat. 52·24, long. 29·50, by observation, and 718 miles distant from Valentia, the *Niagara* having laid 130 miles of cáble.

After this the wind freshened, and a heavy swell got up, increasing the motion of the ship very much, and at midnight it was blowing hard from south-south-east, the consumption of coal required to keep up the speed which I desired to maintain being so great that some apprehension was felt in regard to the sufficiency of our supply of fuel.

At noon on the 31st the *Agamemnon* had paid out 280 miles, and the *Niagara* 285.

The weather did not allow of any observation, but our run by dead reckoning made us about 605 miles from Valentia, and in the locality where the depth of 2,400 fathoms (the greatest in our route) was obtained by Captain Dayman, in Her Majesty's ship *Cyclops*, last year.

During the day the wind continued to blow heavily, the sea running very high. By midnight the barometer had fallen to 29·50, and everything indicated a change for worse, rather than for better, weather. We had then paid out 358 miles of cable, the *Niagara* 365.

At noon on Sunday, August 1st, we were 478½ miles from Valentia, our position by observation being lat. 52° 26′ 30″, long. 23° 16′ 30″, 434 miles having been paid out from the *Agamemnon*, and 440 by the *Niagara*.

During the morning the wind had changed to the south-west, and the weather gave signs of amendment ; but a heavy swell remained, and in the afternoon the breeze freshened, squalls followed each other in rapid succession, and the ship pitched as much as before.

By noon of the 22nd we were in lat. 52·35, long. 19·48, 351·0 miles from Valentia, 605 miles of cable having been laid from the *Agamemnon*, and 615 from the *Niagara*.

In the afternoon the force of the wind decreased, and the motion of the ship was much easier. At 3 p.m. we had to alter our course for a few minutes to avoid a three-masted schooner, which passed us on the port bow so closely as to make it a subject for congratulation that she did not cross our path astern. The cable grew out very much to the starboard side during the change ; but I caused an additional amount of slack to be paid out at the time, so that no undue strain came upon it.

During the evening the weather was squally, and by 4 o'clock in the morning of the 24th the wind had got round to the north-west, and a long slow swell from the south-west caused the ship to pitch and roll as much as before. At this time some excitement was created by a bark bearing down upon our starboard beam ; we increased speed to clear her, but she hove to on being intercepted by the *Valorous*.

APPENDIX

At noon on the 3rd we had paid out 776 miles of cable, being then in lat. 52° 26', long. 16° 7' 40", 212·2 miles from Valentia, the *Niagara* having laid 780 miles.

After the depth of water, which had averaged 2,000 fathoms since the 1st inst., began to lessen ; and at 5 p.m. the greatest variation in our track (from 1,750 to 550 fathoms within about 10 miles) occurred, an extra percentage of slack being laid to provide for any irregularities which might there exist in the bottom. By midnight the depth had further decreased to 216 fathoms.

At 4 a.m. on the 4th the large coil in the main hold was exhausted, and we commenced paying out from the upper deck coil.

By noon the water had deepened again to 400 fathoms ; we were then in lat. 52·11, long. 12·40, only 89½ miles from Valentia, having laid 924 miles of cable, while the *Niagara* had laid 925.

During the day the wind and sea dropped ; and at 8 p.m., having reduced our distance from Valentia to fifty miles, the *Valorous* steamed ahead to make out the land.

The water now shoaled gradually. At 8.30 p.m., having finished the second coil, a change was effected to the cable on the orlop deck.

At midnight we were in company with the *Valorous* in sight of the Upper Skellig light ; and at dawn on the morning of the 5th, abreast of the Blasquets, steaming slowly towards Valencia.

At 6 a.m. we anchored in Doulas Bay, 2,022 nautical miles having been paid out between the two ships, and proceeded to coil a sufficient length of cable to reach the shore into one of the paddle-box boats of the *Valorous*.

The wind freshened in the course of the morning, by which the landing of the end was somewhat delayed, the swell becoming so great that Captain Preedy got up steam in the *Agamemnon*, ready to put out to sea at any moment.

At 3 p.m. the end of the cable was safely brought to the beach, and passed into the Company's station.

The strain upon the cable varied during the paying out under different circumstances of weather, depth of water, and speed of ship, as will be seen by the accompanying tabular log, which furnishes details recorded several times in each hour of the indicated strain, weight on breaks, angle of cable, rate of paying out, rate of ship, revolutions of screw, distance run according to Massey's log, distance made good by observations, and a journal of all the events worthy of note in each watch. An entry is also made of Greenwich time, so that the electrician's diary, and the log kept on board the *Niagara*, may be more readily compared with it.

Some inconvenience was experienced by the great accumulation of pitch and tar, a second coating of which was laid on the cable when coiled away at Keyham for the winter to prevent it from rusting ; but

APPENDIX

this had also its advantage in keeping down the cable leading from the coil, which had, if too dry in any place, a tendency to fly out when running at a high speed.

The paying-out machinery has worked exceedingly well.

The handwheel, for lifting the weights when required, was of considerable service during the unfavourable weather which prevailed for the chief part of the voyage.

The amount of slack paid out amounted to 17 per cent. upon the distance run. Less might have been laid, but I considered it desirable to insure the cable laying everywhere on the bottom—that ample slack should be used to cover any irregularities within bounds of probability.

I must not conclude this report without again expressing my deep sense and appreciation of the laborious zeal and untiring patience exhibited by Captain Preedy and the officers and company of the *Agamemnon* ; nor can I too strongly express my obligation to Mr. Canning and Mr. Clifford, who so ably took part with me in the general superintendence of the work ; and to Mr. Hoar and Mr. Moore, whose supervision of the dynamometer and machinery was of the utmost value to us ; and it must not be forgotten that Captain Hudson and the officers and crew of the *Niagara*, with Mr. Everett and Mr. Woodhouse, who had charge of the operation of paying out from the *Niagara*, with the assistance of Mr. Kell, have also performed their share of the labour equally with those who have returned to Ireland in the *Agamemnon*.

I have the honour to remain

Your most obedient servant,

CHARLES T. BRIGHT, Engineer-in-Chief.

22, OLD BROAD STREET, *August 19th*.

APPENDIX IX*a*

FIRST ATLANTIC CABLE

CURRENTS observed during 1857–8 expeditions by Mr. H. A. Moriarty, R.N.[1]

(*a*) As the current observed is generally between two observed positions both are given.

(*b*) Where no time is specified the current is supposed to have been during the previous twenty-four hours.

(*c*) On many days there was no current, therefore the average strength would would be very low—perhaps half a knot—and direction variable ; that is, in the centre of the Atlantic, where the cable was joined in 1858.

H.M.S. *Agamemnon.*

1857.						
Aug. 10.	Lat.	52·26 N.	Long.	16·0	W.	
						Current N. 32 W.–9½ miles.
,, 11.	,,	52·34	,,	17·39		
						,, N. 58 W.–10 miles.
,, 12.	,,	52·41	,,	15·16		
						,, N. 33 W.–7½ miles.
,, 13.	,,	50·38	,,	10·24		
1858.						
June 10.	,,	49·15	,,	6·44		
						,, N. 61 W.–10½ miles.
,, 11.	,,	49·43	,,	10·8		
						,, North–12 miles.
,, 12.	,,	50·10	,,	13·26		
						,, N. 65 W.–11 miles.
,, 13.	,,	50·18	,,	16·58		
						,, N. 18 W.–19 miles.
,, 14.	,,	51·25	,,	18·55		
						,, N. 77 W.–10 miles. (16 to 20, heavy gale).
,, 15.	,,	52·5	,,	21·1		
,, 21.	,,	54·41	,,	27·28		
						,, N. 50 E.–9½ miles.
,, 22.	,,	53·27	,,	29·49		

---

[1] Now Captain Henry Moriarty, R.N., C.B.

APPENDIX

June 25.	Lat.	51·54		Long.	32·33		

Current N. 38 E. 16½ miles.

" 26. " 52·24 " 31·59

" During 4 previous days, N. 6° E.–51 mls.

July 3. about 52·20 & " 33·34

" N. 66 E.–14 miles.

" 4. Lat. 52·5 " 33·0

" N. 40 E.–10 miles.

" 5. " 52·20 " 33·47

" in 48 hours—E. 24 miles.

" 7. " 51·48 " 26·3

" None.

" 8. " 51·40 " 21·53

" E.–6½ miles.

" 9. " 51·37 " 16·54

THE day after the successful completion of the first Atlantic cable (1858) *The Times* contained the following leader :—

By a chain of electric communication, extending from Trinity Bay, Newfoundland, to this metropolis, we are informed that the last attempt to lay the Atlantic Telegraph has succeeded, and that the Old and New World are actually linked together by the magnetic wire. Although the weather was unfavourable the cable seems to have paid out with the greatest regularity, the quantity of cable discharged from the two ships being the same every day within ten miles.

We sincerely congra ulate the promoters of this great enterprise upon the triumphant success by which, after so many delays and disappointments, they have been rewarded. It is difficult so suddenly to realise the magnitude of the event which has just taken place ; the accomplishment of this mighty feat comes upon us not in the gradual and tentative manner in which most scientific exploits have been performed, but with a rapidity worthy of the agent which it employs. The steam engine, the other great discovery of our time, has been perfected little by little, and no one can exactly say when it was that each of the triumphs which it has successively achieved became possible. Practice was so far ahead of theory that high scientific authorities argued strongly against the possibility of results, and were not refuted by counter-arguments, but by the accomplishment of those very results the possibility of which they had denied. With the Atlantic Telegraph it has been just the contrary. Instead of proceeding by slow degrees, the projectors have leapt at once to a gigantic success. We believe we are correct in stating that 500 miles of telegraph have never before been successfully laid under water, and yesterday we received intelligence that a communication is fully established beneath 2,000 miles of stormy ocean, under a superincumbent mass of water the depth of which may be calculated in miles. Only now, when it has succeeded, are we able fully to realise the magnitude and the hardihood of the enterprise. Over what jagged mountain ranges is that slender thread folded ; in what deep oceanic valleys does it rest, when the flash which carries the thought of man from one continent to another darts along the wire ;

APPENDIX

through what strange and unknown regions, among things how un-couth and wild, must it thread its way!

Since the discovery of COLUMBUS nothing has been done in any degree comparable to the vast enlargement which has thus been given to the sphere of human activity. We may, now that this the most difficult problem of all has been solved, be justified in anticipating that there is no portion of the earth's surface which may not be placed in immediate communication with us. We now know that we have in our hands the means of a practical ubiquity. Distance as a ground of uncertainty will be eliminated from the calculation of the statesman and the merchant. It is no violent presumption to suppose that within a very short period we shall be able to present to our readers every morning intelligence of what happened the day before in every quarter of the globe. More was done yesterday for the consolidation of our empire than the wisdom of our statesmen, the liberality of our legislature, or the loyalty of our colonists could ever have effected. Distance between Canada and England is annihilated. For the purposes of mutual communication and of good understanding the Atlantic is dried up, and we become in reality as well as in wish one country. Nor can any one regard with indifference the position in which the Atlantic Telegraph has placed us in regard to the great American Republic. It has half undone the Declaration of 1775, and gone far to make us once again, in spite of ourselves, one people. To the ties of a common blood, language, and religion, to the intimate association of business and a complete sympathy on so many subjects, is now added the faculty of instantaneous communication, which must give to all these tendencies to unity an intensity which they never before could possess.

Let those who are assembled at Cherbourg to celebrate another development in the art of destruction, and to *fête* the inauguration of a fortress avowedly designed to threaten the independence and pros-perity of these islands, reflect on the true nature of the enterprise which has thus been executed, and turn from the contemplation of science degraded into the handmaiden of slaughter and devastation to science applied to their legitimate office, as the conciliator, the bene-factress, and the enlightener of the whole human race.

APPENDIX X

On his return in H.M.S. *Bulldog* from sounding and exploring the North Atlantic for the route of the proposed cable, Captain (now Admiral) Sir Leopold M'Clintock, in the course of a letter, reported to Sir Charles Bright as follows :—

PORTSMOUTH, 6th Dec., 1860.

MY DEAR SIR CHARLES,—

You are very welcome to my opinion respecting the North Atlantic Telegraph route ; it has been formed solely upon what I have recently seen, and upon the local information I have recently gleaned whilst employed in sounding and surveying along the projected route in command of H.M.S. *Bulldog*.

As for the short lengths of cable between Scotland and Farõe, and from thence to the east shore of Iceland, no difficulties need be encountered ; there are certain channels between the Farõe Isles where the tides are very strong, but there are also still-water creeks, and these, I presume, will be selected for landing the shore-ends.

Like most of my countrymen, I was profoundly ignorant of the climatic condition of Iceland before I went there ; and being desirous of obtaining authentic information respecting the ice movements in the adjacent seas, a scientific gentleman to whom I had an introduction, kindly met my wishes by hunting up for me such facts as I required from the celebrated Icelandic Annals, which date back as far as the ninth century. Briefly, then, the shores of Iceland are only visited by drift-ice about seven or eight times in each century, and it is only upon two or three occasions that the drift-ice is sufficiently extensive to reach the south side and surround the whole island. True icebergs are never seen ; the heavy masses often so called, are small enough to float freely in comparatively shallow water, so that a cable would remain undisturbed at the bottom, the shore end being carried up a fiord.

I trust you will adhere to the original intention of a land line across Iceland to Faxe Bay, as by so doing you will avoid the only part of the sea where submarine volcanic disturbance may be suspected. Faxe Bay never freezes over, and I find no record of drift-ice within it since 1683. Merchant vessels come and go throughout the winter.

From here the cable to Greenland will proceed. During the spring

482

APPENDIX

of the year South Greenland is usually enveloped with drift-ice and icebergs, whilst in the autumn it is for the most part free : it is however, liable to be more or less obstructed throughout the whole year : the present, for instance, has been an extremely bad one—probably the worst for thirty years, and yet the *Bulldog* (paddle-steamer), the *Fox* (screw), and the Danish sailing vessel, have visited Julianshaab in September, October, and November.

My humble opinion is, that in two years out of every three the ice will be too inconsiderable to obstruct the laying down of a cable during the autumnal months, whilst in my official report to the hydrographer of the navy I have expressed a confident opinion that the shore end may be carried into a fiord, where it will be secure from icebergs or drift-ice.

It only remains for me to notice Labrador : the landing-place for a shore end there has yet to be sought for ; the interior, I was told, would present no difficulty whatever for a land line.

The contour of the sea bottom and depth of the ocean throughout the route are decidedly favourable.

You cannot have failed to derive much encouragement from the explorations of the *Bulldog* and *Fox*. There are difficulties, of course, in so great an undertaking, but they are not by any means insurmountable. To me it will always prove a source of real gratification to have been actively employed in the advancement of an object of such vast national importance.

You must take my opinion for no more than it is worth : you are well aware of its chief merit—an entire absence of prejudice or bias of any kind.

Should you desire it, I will be happy to explain to you in detail my grounds for every opinion, and authority for every fact which I have stated.

   I am, my dear Sir Charles, faithfully yours,
       F. LEOPOLD M'CLINTOCK.

To Sir Charles Bright, etc., etc.

APPENDIX XI

THE NORTH ATLANTIC TELEGRAPH

SHORTLY after the return, in December, 1860, of H.M.S. *Bulldog* and the steam yacht *Fox* from exploring the North Atlantic for the proposed cable, the *Mechanics' Magazine* printed the following leading article, which will be found to give a good account of the work done : —

We are much gratified in being able to announce that the North Atlantic Telegraph is about to be proceeded with under the auspices of its original promoters, whose hands have fortunately been strengthened for the work by the devotion of Sir Charles Bright's great abilities and experience to the undertaking. The presence of an English electrical engineer of recognised eminence in the midst of the enterprising gentlemen who have entered upon the realization of this great project is precisely the one thing that has been felt essential to the public approval of their plans ; and we do not hesitate to say that of all the electrical engineers in the kingdom, Sir Charles Bright is best capable of conducting such an undertaking to a successful issue. Those who are intimately cognisant of the circumstances under which the old Atlantic cable was laid, are well aware that it was to the skill, the energy, and the masterly management of Sir Charles Bright, more than to any other causes, that the splendid, though but temporary, success of that undertaking was due. The Atlantic cable "went to the bad" it is true ; but it was laid—it was telegraphed through—it served useful purposes—it thrilled two worlds with fraternal greetings —it solved great problems in electric telegraphy, and it kindled another glorious light on the path of human progress. For all this, we repeat, we have chiefly to thank Sir Charles Bright, whose cool, and wise, and tireless devotion to his great work marked him as the fit recipient of those knightly honours which the Sovereign at once conferred upon him. Moreover, the experience gained in connection with that enterprise was observed more closely, and treasured up more carefully by him than by any other person, and hence he is best able to turn it to good account. For all these reasons we are glad to know that Sir Charles has turned earnestly to the working out of the great problem under the new circumstances.

APPENDIX

Since we last wrote upon this subject much additional information, of the highest importance, has come into our hands. In the first place we have a letter addressed to Sir Charles Bright by Captain Sir Leopold M'Clintock, in reply to an inquiry made by the former gentleman. This letter deserves the most careful attention, because every word proceeding from a man of such pure integrity of character as Sir Leopold is far above the reach of all suspicion as to its good faith, and also because of the just eminence of the writer as an authority on such a question as that under consideration.

The next letter which we propose to notice is from no less a person than the celebrated Dr. Rae—a man whose honourable fame lends weight to every word he pens, and whose experience in northern regions has been too great, and is too well known to need mention here. Dr. Rae, as most of our readers know, joined the late expedition in the *Fox* from a pure desire to aid in the great international undertaking in furtherance of which it was despatched, and engaged to take charge of the land portion of the expedition for the examination of the North Atlantic telegraph route. His full and detailed report is already drawn up ; but as he is awaiting the completion of the enlarged maps of the respective lands traversed—which have been applied for since his return, and are in progress of construction—he has, for the present, submitted to Mr. Croskey (who nobly undertook to bear the costs of the expedition) only a summary of the results of his observations, and of such reliable information collected by him *en route* as may in future be of use to the enterprise. The first across country exploration undertaken consisted in travelling over the island of Stromoe, the largest of the Faröe Islands, in a north-westerly direction from Thorshaven to Haldervig—a distance, by the route followed, of about twenty-seven miles. The journey occupied three days ; and the results were very satisfactory, showing, Dr. Rae says, "that there would be no great difficulty in carrying a telegraph wire over the line of route by which we passed. Two small bays at the termini of the land journey seem well fitted for the landing of a cable ; and the Sound which separates Stromoe from Osteroe offers great facilities for the transport of materials." The mode of transport overland is, he tells us, by pack-ponies, which are extremely sure-footed, and can carry a load of 200 pounds or more. Footpaths already exist which could be much improved at a very moderate expense. The inhabitants of the Faröes are hospitable, well educated, and intelligent, and would give all the aid in their power to carry out this great undertaking, in which they take a deep interest. This information seems quite conclusive as to the practicability of carrying the telegraph across the Faröes if desirable.

Coming now to Iceland, the subject assumes increased importance from a passage in the letter of Captain M'Clintock. " I trust you will

APPENDIX

adhere," he says, " to the original intention of a land line across Iceland to Faxe Bay, as by so doing you will avoid the only part of the sea where submarine volcanic disturbance may be suspected." Let us now hear what Dr. Rae says on the practicability of such a land line :—

The journey across Iceland [he says] from Benfiord on the east, to Reikjavik on the west coast, commenced on the 15th August, and occupied fourteen days, the distances travelled over being little, if at all, short of 450 miles. During this journey, although we by no means followed the shortest route, or that best adapted for telegraphic purposes, no obstacles, as far as I could notice, presented themselves to the erection of a telegraphic wire. As we kept well to the northward all the streams of any size crossed by us (with one exception) flowed in that direction, and high, well-defined banks, which showed no indication of ice action, or of shifting their position, as I am told, is a peculiarity of the rivers flowing into the sea on the south coast of the island.

By information received from some of the most intelligent and respectable gentlemen at Reikjavik, I am led to believe that Portland on the south coast of Iceland, although not a harbour for vessels, is well fitted for the landing of a cable, the beach being of fine sand. There is frequently a heavy surf, but this, I am told, would not injure a strong cable laid upon a shore of the nature described.

The modes of transport of stores over Iceland are by ponies, and in some places wagons and sledges may be used.

The lowest temperature experienced during winter is 13° or 14° below zero of Fahrenheit, and this is of rare occurrence.

With regard to the visit of the *Fox* to Greenland, the experience gained in reference to the land part of an expedition was very limited, in consequence of the lateness and unusual severity of the season and other causes. Dr. Rae states that he had no opportunity of examining the inland ice to an extent, so that he cannot give an opinion on the practicability of carrying a wire or cable across any part of South Greenland. But this circumstance is, as he says, of little importance, as no one now believes that it will be found requisite or advantageous to construct a land line there. It is important, however, to know that, from his own observation and from information obtained from the most reliable sources, Dr. Rae does not believe that the ice, either in the form of floes or bergs, can injure a cable if once down, and that in ordinary seasons a cable may be laid to and from Greenland without much difficulty. That the dangers to ships from ice are not generally considered very great in the South Greenland seas may be inferred, he says, from the fact that vessels, sometimes numbering ten or twelve in the season, visit this coast, and generally return laden to Europe in perfect safety and uninjured. Most of these vessels are not strengthened in any way to protect them from ice.

We think we have now said enough to show that the promoters of

APPENDIX

this North Atlantic Telegraph have good grounds for anticipating success telegraphically, and therefore good grounds for coming before the public with a proposal to construct such a telegraph. There is no need to undervalue the difficulties that have to be encountered ; they are numerous, but, as Sir Leopold M'Clintock says, they are not by any means insurmountable, especially to such men as those who have the enterprise in hand.

We will only add to these citations a further remark or two from a paper which has just been drawn up by Mr. J. W. Tayler, F.G.S., who has just returned from Greenland after a residence of seven years upon and near the west coast. Mr. Tayler says in one place : "Place exist in Greenland at which the cable can be brought to land from depths of water so great as to preclude the possibility of icebergs grounding upon it." And again : "As regards the land of ice of Greenland, there will be experienced no greater difficulties than in laying a cable of equal length in any other part of the globe, provided of course that every prudent precaution be taken in the choice of the proper season and locality for landing the cable in Greenland." It is due to Mr. Tayler to say that, although the paper from which we take these extracts is the first document in which his name appears in con-nection with the new telegraphic enterprise, he has verbally contributed very largely to the knowledge which is now possessed in reference to the Greenland coasts. In fact, nearly all that is known reliably con-cerning them is due to his close acquaintance with the country and his extreme liberality in imparting what he knows to others. We think the promoters of the enterprise would do well to obtain his advice respecting the selection of the fiord into which the cable shall be taken, as any error in this respect may prove very detrimental to them.

APPENDIX XII

THE following is the Paper read by Bright before the Geographical Society on January 28th, 1861 :—

SYNOPSIS OF THE SURVEYS OF THE *FOX*

BY SIR CHARLES T. BRIGHT, F.R.G.S.

I have been requested by the promoters of the North Atlantic Telegraph to present to the Royal Geographical Society a synopsis of the report which has been handed to me by Captain Allen Young, upon his recent voyage in the steam yacht *Fox*, and his careful and elaborate survey of the proposed telegraphic route between Europe and America, by way of the Faröes, Iceland, and Greenland.

FARÖE ISLANDS

This most interesting group of isles, the capital of which is Thorshaven, lies some 200 miles north of Scotland, and is under the authority of the Danish Crown. I will not occupy the time of the Society in discussing the political, physical, or other characteristics of these islands, but proceed at once to quote some interesting extracts from Captain Young's report. He says :—" We were naturally anxious to reach the spot at which our work was to commence, and to ascertain the first foreign station at which the telegraph cable was to be landed. We were glad to make the Faröe Islands, distant 50 miles, on the evening of the 2nd August, the remarkable clearness of the atmosphere and the height of the land making our distance from it apparently far less than we were by our observations. When 46 miles E.S.E. from Naalsöe we obtained soundings in 102 fathoms, sand and shells. We here passed through many patches of discoloured water of a reddish hue, caused probably by minute animalcules on the surface, specimens of which, brought up by the towing-net, were preserved. Specimens of water, both on the surface and at various depths, were frequently obtained during our voyage and preserved, the temperature and specific gravity being registered in the meteorological journal.

" On the morning of the 3rd the land was obscured by clouds and mist, which, as the sun rose, gradually dispersed, and enabled us to obtain views of the land, and also to fix our positions by Born's chart

APPENDIX

to commence a line of soundings into the north point of Naalsöe; the depths were 36 to 26 fathoms, with a bottom of sand and shell."

Thorshaven.—" On rounding the north point of Naalsöe we took a fisherman on board as pilot, and at 10.50 anchored in Thorshaven, and immediately commenced an inquiry and examination of the locality, and testing the accuracy of all the charts and maps in our possession. The results were as follow : Thorshaven and bay is protected by Naalsöe, and is land-locked, excepting on two points to the south-east, and on one point to the north-east. A swell sets in to the inner harbour with south-east gales ; but this cannot be to any very great extent, from the fact that vessels lie at their moorings throughout the winter. The bay has good anchorage, varying in depth from 25 to 8 fathoms, bottom of sand, gravel, and shells, with a few patches of hard ground. Vessels usually moor in the two inner harbours or creeks, the northern being most frequented for the facility it offers for loading and discharging cargoes. Either of the inner harbours would do very well to land the telegraph-wires, but from the many vessels frequenting the port it appeared desirable to select another place, for even were the cable to be buoyed, the risk from the ships' anchors would be considerable, on account of the want of space ; but half a mile southward of Thorshaven is a small cove called Sandygerde, where the cable could be landed in safety and clear of ships' anchors. This cove is 1½ cable's length across and about the same depth, and shoals gradually to a sandy beach ; it is intersected at the head by a watercourse and mill, the land sloping gradually up an extensive valley to the interior. As many additional soundings were obtained across the fiord as our time would admit, proving that, although the channel is uneven, there is nothing to prevent bringing a cable in from sea. From the most reliable information from pilots and our own observations, the stream on the flood never exceeds 4 knots on the strongest spring tides, whilst on the ebb it is much weaker, and at times scarcely perceptible. It is high water at full and change at 4 o'clock : the flood runs to the southward. The Gulf Stream appears to sweep round these islands from left to right, or direct as the hands of a watch ; and therefore in sailing from Thorshaven for the northward, by starting with the first of the flood and passing to the southward of Stromöe, and through Hestöe and Westmanshaven Fiords, you can carry a 9 hours' favourable tide. The rise of tide at Thorshaven does not exceed 6 feet."

Westmanshaven.—" We left Thorshaven at 1 p.m., passing through Hestöe and Westmanshaven Fiords, and anchored in Westmanshaven in the evening. The scenery in these fiords is very magnificent, and as we steamed through with a strong head-wind and weather tide, the surface of the water covered with sea-birds, the lofty hills on either hand rising to the height of 1,500 to 2,000 feet, with vast

489

APPENDIX

basaltic caverns and columns in the cliffs, formed a picture not easily forgotten. As I had heard that Sir Leopold M'Clintock had already examined this port, I did not deem it necessary to delay the ship for that purpose. The fiord appears clean and clear, with deep water close into either shore. I was informed that there is 70 fathoms water in the middle, a little north of Welbestad, and the stream in strongest spring tides runs 6 knots through the fiord. The rise of water is much influenced by the winds outside ; it has reached 10 feet at spring tides, and has been known as low as 4, but the mean rise appears to be from 6 to 8 feet. Westmanshaven is said to be the best harbour in the islands ; it is completely landlocked, with a bar, probably formed of the *débris* washed down from the surrounding hills, and accumulated by the action of the streams and eddies in the fiord. I fear the current in this fiord would be disadvantageous."

Haldervig.—" We left Westmanshaven on the evening of August 5th, and, after weathering the northern extremity of Stromöe, entered the sound between Stromöe and Osteröe, and anchored at Haldervig at 11.30 p.m. on the same night. An examination of the port and estuary of the sound was commenced. The results of these observations, which occupied two days, were, that little or no stream is found in the sound ; that Haldervig has good anchorage, and is perfectly landlocked ; the deepest water is 34 fathoms, bottom black mud and sand, but that a sand-bar exists between Eide Point and Stromöe, over which there are 8¼ fathoms in the deepest part ; and as in northerly gales the sea is said to break upon it, I consider that the cable would require a strong shore end to ensure its safety in crossing this place. This bar lies rather within the entrance and narrowest neck of the sound." In the summary Captain Young states :—" At Haldervig we surveyed harbour and fiord, and found all satisfactory, and I think that place to be well adapted for the reception of the cable. We found but little current, and the cable can be taken in, in a tolerable depth of water, into a perfectly land-locked position."

ICELAND

" On approaching the coast of Iceland we got occasional soundings towards Ostrê Horn, under which we were obliged to anchor in a dense fog, after getting into an extensive and dangerous reef of rocks, called by the Icelanders the Hartinger and Bortinger. These reefs lie two miles east (true) off this cape. They do not appear on the Danish surveys, but I afterwards found them as a single rock upon a French chart."

Beru Fiord.—" On the morning of August 12th, the fog having lifted, we weighed under steam, and got into a position to carry a line of soundings into Beru Fiord, between the islands of Papey and Kogar Point ; these soundings average about 30 fathoms, principally sand and shells. We anchored off Djupivogr factory the same day, and it being Sunday we ceased operations during the afternoon. The weather that day was the finest we had had since we left England, and the evening was truly summer-like. During the following five days, and when not prevented by the prevalent rain and fogs, we proceeded with the examination of the fiord, and finding it would not be advisable to carry a cable into the small harbour of Djupivogr, on account of many rocks in its vicinity and its being the anchorage of the small vessels frequenting the coast, we sought for a more suitable landing-place higher up the fiord, and succeeded in finding an excellent bay, called Gautavik, on the north shore, five miles from the entrance. A depth of nearly 30 fathoms can be carried in from sea to within a quarter of a mile of the shore, while the bay itself afforded good protection and anchorage for any large ships that might be employed in the undertaking.

" High water at Djupivogr at 3 o'clock, full and change ; rise 6 feet. The tide has been known to rise 6½ feet before the coming of easterly gales ; about the same time flood outside runs S.S.W. (true), ebb N.N.E., between Papey Island and the main. The strongest known stream has 4 knots, but the average in ordinary spring tides is not more than 2½ knots.

" In 1860 drift-ice appeared off the coast and entered the fiord, and again (though in very small quantities) in 1859-60 (*sic*). This ice, called here Greenland ice, is the ordinary washed and decayed floe-ice, and comes from the north-west. *No icebergs have ever been seen on the coast.* The drift-ice appears with northerly and departs with southerly winds, and less of it comes into Beru Fiord than any other fiord on the east coast of Iceland, the residents accounting for this fact by Beru Fiord having a south-west direction, and is consequently protected by the more northerly and projecting capes which shunt the ice off, while the local tides keep it drifting up and down the coast. The fiord itself never freezes, but thin ice has been known to cover the harbour off the factory for a day or two during the winter.

" A tolerably complete survey of the fiord from the entrance to Gautavik was completed, but a further examination would be advisable outside, to ascertain the proper channel in which to lay the cable. The greatest difficulties experienced on the coast by seamen are from the prevalent fogs during the summer months, and with easterly winds, *and this would render it advisable to start from this coast towards Faröes*, in laying the cable, because making a good landfall here would be attended with considerable uncertainty." Finally, as to the practicability of Beru Fiord, Captain Young says : " There will be no difficulties from the sea, ice, or otherwise, and the only obstacles will be from fogs and thick weather, but which may be overcome by

APPENDIX

selecting proper seasons, and taking precautions in landing or embarking the telegraph cable."

Reikjavik.—Captain Young sailed from Beru Fiord on the 17th day of August, and arrived at Reikjavik, the capital of Iceland, on the 21st day of August, and, after making inquiries as to the coasts, he says : "I then determined to examine Hval Fiord, as from its situation it appeared to have the advantage over any place in Faxe Bay, and on the 27th I proceeded up that fiord, sounding it as far as Maria Havn, a small harbour and salmon river on the south shore, 7 miles from the entrance of the fiord. The least depth of water in the channel of the fiord is 14 fathoms, with deeper water both outside and in, the general depth being 18 to 20 fathoms, soft mud. The cable could be taken into Maria Havn through soft mud, on a sandy beach in a landlocked position. Hval Fiord is protected from a heavy sea breaking into it by the shoals of Vestrhram and Sydiahraun, in Faxe Bay, and on which there is less water than in the shoalest part of the channel of the fiord. The bays in the fiord are sometimes covered with thin ice, but the fiord itself never freezes ; and with reference to drift-ice on this part of the coast, I cannot do better than quote the words of Sir Leopold M'Clintock : 'Faxe Bay never freezes over, and I can find no record of drift-ice within since 1863. Merchant vessels come and go throughout the winter.' "

GREENLAND

The *Fox* left Reikjavik August 31st, and after a very rough passage arrived at Frederikshaab, October 2nd. Captain Young remained there to make some necessary repairs, and finally arrived at Julianshaab on the 22nd October. He then reports : "Having made all inquiries about Igalikko or Julianshaab Fiord, I deemed it advisable at once to commence a survey of this beautiful arm of the sea, and acting upon the opinion of Colonel Shaffner, that were this fiord found practicable, the electric circuit from Reikjavik would not be too extended.

Julianshaab Fiord.—"We first sounded up to the head of the fiord, which gave an opportunity for our landing a travelling party, under command of Dr. Rae, to examine the inland ice and nature of the country. A party also went to the Old Nordisker Ruins at Igalikko." Returning with the *Fox* to Julianshaab, October 27th, Captain Young then surveyed the estuary of the fiord, and from the soundings obtained says : "I am of a decided opinion that *a depth of not less than* 150 *to* 160 *fathoms* can be carried from the middle of the fiord abreast the settlement out to sea, with a general muddy bottom.

"*The depth of water will effectually preclude injury to the cable from the largest icebergs ever seen upon the coast.* Although many bergs lay along the coast, we saw none aground in this valley of the fiord, nor, according to information obtained from the residents, have they been seen grounded in that channel." Captain Young then proceeds to say : " This report and my previous letters will show that my decided opinion (so far as we have been upon that route) is favourable to the practicability of the undertaking, and that Julianshaab will, under all circumstances, be well adapted for the reception of the cable. With regard to the operation of laying the cable, I consider that no apprehension may be felt on that point, for, from the sudden disappearance which we witnessed of the ice from the coast, and from the ice *usually* dispersing from the south-east shores of Greenland in the autumnal months, opportunities will always occur when a ship having the cable on board, and lying in readiness in Julianshaab, may depend upon having a period of clear and open sea. *The cable once laid, no drift-ice can in any way injure it, if the proper precautions are taken in securing the shore end.*

Ice of the Greenland Seas. —" Since my arrival I have seen the admirable remarks of Mr. J. W. Tayler upon the southern coast of Greenland, the results of his experience during seven years' residence there. His opinions must be most satisfactory to you, and I am sure that all who are interested in the work must be grateful to him for having so freely given them.

" I perfectly coincide with his views with regard to the size of the icebergs frequenting the above coast and accompanying the Spitzbergen drift-ice ; and as this bears upon my own opinion, that no iceberg will ground in the channel of Julianshaab Fiord, I think I may here explain my reasons for this statement. Having navigated the entire west coast of Greenland, and into all the principal settlements, and having experienced a whole winter's drift in the ice, through Baffin's Sea and Davis' Strait, I have had occasion to remark and to gather all possible information upon the ice movements.

"Around the coast of Greenland, westward of Cape Farewell, there are two distinct descriptions, or rather kinds, of drift-ice ever approaching, but never meeting together. The first is the ice formed during the winter on the vast area of Baffin's Sea and the different channels from the Polar Seas westward of Greenland. This ice, called by the Greenlanders *the west ice,* often blocks up throughout the year the upper part of Melville Bay, and drifts constantly throughout the winter and early spring to the southward, through Davis Strait, into the Atlantic. It seldom comes in contact with the coast of Greenland below the parallel of Disko, *and there is always an open sea between it and Greenland as far up as Holsteinberg throughout the winter.* The second is the Spitzbergen, called also the ' store ice,' which, as has been shown, comes down the east coast of Greenland, around Cape Farewell, and is carried by the current up the west

APPENDIX

coast, at times even to the Arctic Circle, but by which time it is usually pretty much broken up, and, if not entirely dispersed, the last remnants are supposed to have returned southward, by Davis' Strait, to the Atlantic—so near these two great ice streams approach, that vessels bound to the colonies have in the early spring passed up Davis' Strait with the west ice and the Spitzbergen ice on either hand. But as there are two kinds of oceanic ice, so also are there two distinct classes of icebergs, namely, the bergs from the stupendous glaciers far up the west coast of Greenland, and especially in Melville Bay ; these bergs attain an astonishing magnitude, but like the west ice, which they accompany or outsail, they do not come upon the west coast of Greenland below the same parallel, although in exceptional seasons of violent gales, such as the last, they may be blown in upon the land a little more to the southward ; and I saw some of these *ice islands* last October aground upon and near Tallert Bank, northward of Fredrikshaab. The other icebergs are those which accompany the Spitzbergen ice, and may be said to follow its movements. They are launched from the glaciers far up the east coast of Greenland, and from those in the island of Spitzbergen ; and besides being originally far less in their dimensions, they are exposed during their long passage southward to the warmer Atlantic winds and heavy swells, and are proportionally reduced before their arrival at Cape Farewell. The bergs from the southern glaciers of Greenland are but small, and need scarcely to be taken into consideration, for, as they must come out from the heads of the fiords, they surely would not take the ground in again entering the *channel* of the deepest fiords.

"With regard to the flotation of ice, it has been calculated that seven-eighths of a cubical mass of ice will be immersed ; but icebergs being very irregular in their formation, and having usually very peaked and angular summits, whilst below the water they are smooth, rounded, and most frequently widened out, I think that icebergs are not found that draw more water than the proportion of six feet below to one foot of perpendicular height above the water. Therefore in 150 fathoms of water (the very least found in the entrance to Julianshaab Fiord) an iceberg of an elevation above the water of 150 feet, or having an entire perpendicular height of 1,050 feet, will there be suspended above the ground, and such bergs are not to be met with in that place."

Remarks upon the Seasons.—" The finest months in the Faröes are June and July, and in these months only should the cable be laid, and then about the last quarter of the moon, because the tides are greater at the full than at the change, consequently the neap tides immediately after the last quarter should be selected, as the currents are then inconsiderable. I have already given my reasons for recommending that the cable be laid from Iceland towards the Faröes, not

APPENDIX

only on account of the prevailing fogs on the east coast of Iceland, but also from the greater facilities for making the coast of the Faröes, and the opportunity that the comparatively speaking shallow water off the north-west coast would give of shipping and buoying the cable in the event of a sudden gale of wind occurring at the time of laying it. The finest months upon the east coast of Iceland are also June and July, but I was informed that the weather is clearer earlier in the season, in the months of May and June—I suppose from the alternations of temperature being then less frequent. A few hours, however, of clear weather would always carry a ship beyond these mists, which usually hang only on the land. With reference to Faxe Bay station, the west coast of Iceland is generally free from fogs, and the Gulf Stream which sets round Cape Reikianess, and appears to keep up a continuous flow around Faxe Bay to the northward, passing out by Snæfellssness, also appears to considerably affect the climatic condition of the west coast. Navigation is open all the year round, and the operation of bringing the cable here can be timed to the opportunities for departing from Greenland. A fine pyramidal beacon has lately been erected on the Skagen, and is of great assistance to navigators entering Faxe Bay from the southward."

Conclusion.—Before concluding, it is proper to state that the voyage was one surrounded with much peril, on account of the succession of gales and the extraordinary quantities of ice found in the Greenland seas ; never within the memory of man has there been so much and so long a continuation of ice upon the Greenland coasts as during the past year. In the arduous labours of the voyage Captain Young was most ably assisted by Mr. J. E. Davis, Master in the Royal Navy, who by the kindness of Captain Washington, hydrographer to the Admiralty, was permitted to accompany the expedition and take part in the necessary surveys ; and his former well-known services under Sir James Ross in the Antarctic regions, and great experience as a marine surveyor, enabled him to render the most valuable assistance in the especial mission of the *Fox*, which is acknowledged by Captain Young in his report in the highest possible terms. During the voyage various specimens of deep interest to the geologist and naturalist were collected ; a large number of scientific observations were made, and a detailed meteorological journal was kept, which, together with other valuable information and an extensive collection of photographs, made with great zeal by Mr. Woods, under very difficult circumstances, have been furnished to the promoters of the enterprise, with the hope that they will be found to contribute to the cause of science, as well as to the immediate object for which they were made. Time will not now permit me to give further details of this most interesting voyage, but any members of this Society who may desire to make personal inspection of the charts, meteorological

APPENDIX

tables, logs, reports, and specimens, will be gladly permitted to do so.

Having thus presented to the Society some of the most valuable and interesting portions of Captain Young's report, I have only to observe that the result of the recent survey has been to remove from my mind the apprehensions which I previously entertained, in common with many others, as to the extent and character of the difficulties to be overcome in carrying a line of telegraph to America by the northern route.

Prior to the dispatch of the surveying expedition we had no knowledge of the depth of the seas to be crossed, with the exception of the few soundings obtained by Colonel Shaffner in 1859, and our information as to the nature of the shores of Greenland in regard to the requirements for a telegraphic cable was equally small.

These points are of vital consequence to the prospects of the North Atlantic Route, and the survey has placed us in possession of satisfactory particulars respecting them. The soundings taken by Sir Leopold M'Clintock will be a guide in the selection of the most suitable form for the deep-sea lengths of the cables, while the information furnished by Captain Young will direct the construction of the more massive cables to be laid in the inlets of the coast. It is not necessary to determine upon the precise landing places and other points of detail in connection with the enterprise at the present time, but the promoters of the undertaking have received ample encouragement from the survey, and from the testimony of competent and experienced voyagers and sojourners in the countries to which the line is to be carried, to warrant them in proceeding with their labours with renewed vigour and confidence. When they have achieved that success which their perseverance and energy deserve, I am sure they will always gratefully remember that their endeavours at the stage of their operations which is now under discussion would have been very much less productive of good results but for the patriotic foresight of Lord Palmerston in ordering the *Bulldog* on her late successful service; and for the assistance of Sir Leopold M'Clintock, Captain Young, Dr. Rae, and the Commissioners appointed to accompany the *Fox* by the Danish Government, as well as others who took part in the cause, whose patience and devotion to their self-imposed work have been above all praise. Nor can those interested in this important undertaking forget the great help rendered to them by the Royal Geographical Society.

Butler & Tanner, The Selwood Printing Works, Frome, and London.

INDEX

Acts of Parliament.—Telegraph (companies) of 1868, 76 ; Telegraph Purchases Regulations of 1868, 76.

Aerial, telegraph posts, system of, adopted, and its object, 50.

Agamemnon, H.M.S., 173 ; selection of, 145 ; description of, 146 ; sets out for Gravesend to adjust compasses, 171 ; loading of, 233 ; in a storm, 239 ; strain on, by storm, 260 ; returns to Queenstown July 12th, 1858, 274.

Airy, Sir George Biddell, K.C.B., F.R.S., report of, 212.

Albany, s.s., chartered for 1866 expedition, 388, 394.

Aldham, Capt. W. C., 234 ; made C.B., 329.

Allan, Mr. Thomas, engineer United Kingdom Telegraph Co., 75.

Amos, Mr. C. E., M.Inst.C.E., modification of Appold's brake by, 205.

Anderson, Capt. Sir James, commands *Great Eastern*, 1865 expedition, 381.

Andrews, Mr. William, secretary United Kingdom Telegraph Co., 75 ; managing director Indo-European Co., 75.

Anglo-American Telegraph Co., 385 relations with Atlantic Co., 391 (note),

Ansell, Mr. W. T., of old Electric Co., a survivor of early telegraphy, 76.

Appold, Mr. J. G., F.R.S., friction brake by, 204.

Armour, protecting second Atlantic main cable, 377.

Arms, coat of, belonging to Bright family, 11.

Ashburton, Lord, North Atlantic project, 371.

Athenæum, The, correspondence *re* Atlantic cable, 126.

Atlantic Telegraph Company : agreement for formation of, 110 ; registration of, 115 ; first meeting of, 116 ; Charles Bright subscribes to, 117 ; election of board, 121 ; American support, 118 ; negotiations with government, 119 ; first meeting of shareholders, 121 ; Magnetic Co. provides capital for, 117 ; Charles Bright appointed engineer-in-chief, 124 ; Mr. Whitehouse appointed electrician, 124 ; Mr. Field appointed general manager, 124 ; Mr. George Saward appointed secretary, 124 ; increase of capital, 201 ; representatives of 1865 expedition, 382 ; practically amalgamated with Anglo-American Co., 385 ; relations with Anglo Co., 391 (note).

Badsworth : Hall, 2 ; monument in church, 5 ; hunt, 6.

Bagshawe, John, of Hucklow Hall, 9 (note).

Ball, Mr. John, F.R.S., North Atlantic project, 371.

Ball-splice, method described, 287.

Bannercross, the manor of, 1 ; estate of, 9 (note).

Banquets, a series to be gone through, 323 ; Lord Mayor of Dublin's (from *Morning Post*), 325 ; Lord Lieut.'s, to Bright, Killarney, 329 ; Liverpool, return of 1866 expedition, 401.

Barlow, Mr. Peter, F.R.S., 224.

Bartholomew, Mr. E. G., 184.

Beechey, Capt. R. B., R.N., oil painting of *Fox* party by, 371.

Belcher, Admiral Sir Edward, North Atlantic project, 371.

Belgium, Holland and, early cable between, 92 (note).

Benest, Mr. H., special grapnel of, 395 (note).

Bergue, Mr. Charles de, 152.

Berryman, Lieut. O. H., U.S.N., Atlantic sounding, 96.

Bewley, Mr. Henry, 54.

Blair, Mr. (of Dunskey), Anglo-Irish cable, through estate of, 87.

Brassey, Mr. Thomas, M.P., supporter of cable enterprise, 375 ; capital for 1866 expedition, 385.

Brett Brothers, house printing telegraph, 73 (note) ; promotion of cable laying, 78.

Brett, Mr. John Watkins, director of the English and Irish Magnetic Telegraph Co., 49 ; early Mediterranean

INDEX

cables, 92 (note) ; registers Co. for an Atlantic cable, 1845, 95 ; takes £5,000 of shares and bonds in "Newfoundland" Company, 108 ; counsels renewed efforts after disaster of 1858, 280 ; cause of cable injury, views on, 357 ; light Mediterranean cables of, 378 (note).

Bright, Mr. Brailsford, father of Sir Charles Bright, 17.

Bright, the brothers (Sir Charles Tilston and Edward Brailsford), inventions of: Magneto - electric telegraph, improvements in, 67; acoustic telegraph (Bright's bells), 67 ; duplex telegraphy, 74 ; fault-testing apparatus, 351 ; curb keys, 352.

Bright, Mr. Charles, F.R.S.E., son of Sir C. Bright, "The Telegraphs of the United Kingdom," article on, by, 76 ; *Submarine Telegraphs*, work on, 93 (note).

Bright, Sir Charles Tilston : descent from ancient Yorkshire family, 1 ; a Yorkshireman, *pace* Dr. Conan Doyle, 14 ; birth and parentage, 17 ; sporting proclivities, 23 ; Merchant Taylors' School, 28 ; bent of school studies, 34 ; interest in electricity and chemistry and early land telegraphs, 35 ; joins Electric Telegraph Co. at fifteen, 35 ; precocious inventor, 38 (note) ; and original "Electric" Co., 35, 39 ; assistant engineer, British Telegraph Co., 39, 44 ; letter to district superintendents and agents, 40 ; letter to his *fiancée*, 46 ; insulator for aerial telegraph wires, 49 ; engineer-in-chief to English and Irish Magnetic Telegraph Cos., 49 ; shackles, 50 ; telegraph posts, 51 ; translator or repeater, 52 ; marriage of, 59 ; apparatus to check decay of gutta-percha, 72 ; connection with Magnetic & London District Co., 74, 76 ; contemporaries of early land telegraphy, 76 ; Anglo-Irish cable expedition, 1853, 83, 88 ; Atlantic cable, most memorable achievements of, 91 ; experiments carried on by, with his brother, on the wires of the Magnetic Telegraph Co., 100, 101 ; copper conductor for cable recommended by, 124 ; appointed engineer-in-chief to Atlantic Telegraph Co., 124 ; vigilance in superintending manufacture of cable, 138 ; moves to Harrow, 142 ; Bright's cable-laying gear, 151, 153 ; speech at farewell banquet, 165 ; takes up his quarters on *Niagara*, 183 ; description of laying of cable up to time of first breakage, 190 ; arrives at Plymouth, 198 ; letter to Mr. Whitehouse on uncoiling, 198 ; letter to Mr. Kell on uncoiling, 199 ; pro-

ceeds to Valentia to pick up cable, 200 ; modification of Appold's brake by, 205 ; dynamometer invented by, 207 ; approached by numerous inventors, 214 ; letter to Atlantic Telegraph Co. on paying out, 221 ; visits home at Harrow, 1858, 280 ; as engineer-in-chief, expedition of 17th July, 1858, 286 ; Institution of Civil Engineers, obituary notice of, 314 ; message to board of Atlantic Telegraph Co., 318 ; official report of success, 1858, 319 (note) ; banqueted, Ireland, 323, 328 ; tribute to, by Prof. Thomson (Lord Kelvin), 323 (note) ; toast of Cyrus Field, 324, 326 ; knighthood conferred on, 326 ; testimonial from directors of Atlantic Telegraph Co., 326 ; presentation at Court, 326 (note) ; youth of, at time of knighthood, 329 ; return to Harrow Weald, 333 ; views on proper size of conductor, insulation and type of core, 347 ; investigation, collapse of 1858 cable, 351 ; curb keys apparatus, 352 ; the North Atlantic route project, 360 ; technical adviser, North Atlantic telegraph, 365 ; paper on results *Fox* expedition, 368 ; consulted as to second Atlantic cable, 377 ; political interests, 382, 391 ; speech at Liverpool, 404.

Bright, Mr. Edward Brailsford, elder brother of Sir Charles Bright, 17 ; joins Electric Telegraph Co., 35 ; early joint inventions of, 36 ; joins Magnetic Telegraph Co., 39 ; on "Retardation of Electricity through Long Subterranean Wires," 63 ; succeeds his brother as engineer of Magnetic Co., 74 ; with Anglo-Irish expedition, 83.

Bright, Major Henry (Royal Irish Fusiliers, 87th Regt.), uncle of Sir Charles Bright, 9.

Bright, Mr. John, formerly M.P. for Pontefract, first master of Badsworth hunt, 6.

Bright, Col. Sir John, Bart., career of, 2 ; monument to, 5.

Bright, John, of Bannercross, 9 (note).

Bright, Dr. John, uncle of Sir Charles Bright, 10.

Bright, Right Hon. John, M.P., supports Atlantic cable on economic grounds, 374.

Bright, Lady, on board during Anglo-Irish cable laying, 80 ; telegram and congratulations to on laying of Atlantic cable, 320.

Bright, Mary, granddaughter of John Bright, M.P., 6, 9.

Bright, Rev. Mynors, Principal of Magdalene College, Cambridge, 10.

498

INDEX

Bright, Lieut. Thomas, R.N., uncle of Sir Charles Bright, 10.

Bright, Mr. William, uncle of Sir C. Bright, 10.

Bright, Rev. William, cousin of Sir C. Bright, 10.

British Electric Telegraph Co., later the "Electric," 39 (note).

British and Irish Magnetic Telegraph Co., the amalgamation of earlier companies, 73. (See under "Magnetic" Co.)

British Telegraph Co., formation of, 39; Charles Bright, letter on prospects of, 40; capital of and shareholders in, 43; European and American Telegraph, combination with, 73; absorbed by Magnetic Co., 73.

Britannica Encyclopædia, article in, on electric telegraph, 49 (note); 359 (note).

Brooke, Lieut. J. M., U.S.N., ocean sounding apparatus, 96.

Brooking, Mr. T. H., 121; resigns vice-chairmanship Atlantic Telegraph Co., 1858, 279.

Brooks, Mr. C. H., submarine telegraphy paper, 211.

Brown, Lenox & Co., Messrs., experimenting at works of, 130; grappling apparatus, 1866 expedition, 389.

Brown, Mr., M.P. (Sir William, Bart.), first chairman Atlantic Co., 121; views of after the failure of 1858, 279.

Brunel, Mr. Isambard Kingdom, F.R.S., and the *Great Eastern*, 224, 379.

Buchanan, President, Letter from, 167; message to Queen Victoria by cable, 1858, 337.

Bull Arm, cable end at Newfoundland, 1858, 318 (313 note).

Bulldog, H.M.S., *Fox* expedition, 367.

Buoys, used in 1866 expedition, 388.

Burton, Lord (Mr. Michael Bass, M.P.), 48.

Cable, Anglo-Irish, 1853, cables which preceded, 78; manufacture of cable and core, 80; cable for, 80; start of, 83; laid, 87.

Cables (early), prior to Anglo-Irish cable of 1853, 78; English Channel, 78, 92 (note); England-France, 78; Black Sea, 78; Dover-Ostend, 79.

Cable, first Atlantic, the achievement of Bright's career, 91; early projects, various lines, 92 (note); difficulties to be overcome, 95; conditions for landing, 102; construction of, 128, 130; iron sheathing for, 128 (note); weight of, 129; length of, 130; conductor, 128; machinery for regulating egress of, 151; drums for, 151; landing at Ballycarberry, 177; description of laying by Bright up to time of first breakage, 190; uncoiling, 199; picked up at Valentia, 200; arrangements for stowing, 224; experiments with at Keyham Dockyard, 235; decision of board to pay out in mid-ocean, 229; experiments with, 236; in a storm, 239 (note); in a tangle after storm, 261; lowering of, 262; breaks, 262; loss of, 264; breaks, 270; damage by storm to, 270 (note); strain on, controlled, 291; injury reported in, 292; splicing against time, 294; breaking strain approached, 294; apparent failure and variation of insulation explained, 295 (note); as "a silver thread among the waves," 300; effect of submersion in deep water on insulation of gutta-percha, 306; first message through entire length at bottom of sea, 313; perfect condition of, after laying, 334; effect of submersion, 334; taken in charge by Mr. Whitehouse, 334; early difficulties in working, 335; early messages through, 313, 335, 336, 337; further early messages transmitted, 340, 342, 343; facsimile of one of first messages received, 341; first public *news* message, 341; "repaid its cost by value of a single message," 341; Indian Mutiny news, peace with China, etc., 342, 343; military messages, enormous saving effected through, 343; failure of insulation, 344; causes of total failure after three months, 344, 345; true cause of catastrophe, etc., 359.

Cable, second Atlantic: cost and type of, 377, 386; length of 1866 expedition, 393; method of recovery of, 395; completion of second line, 395; length of line commenced, 1865, 399; rate of working at first, 400; temporary breakdown of, 401.

Cables, underground system of, 54.

Canning, Mr. (Sir Samuel), chief assistant, 134; repair of cable, 292; laying of cable end, 1858, 313; machinery fitting, *Great Eastern*, 380; engineer in charge, 381; with 1866 expedition, 380.

Cape Breton cable, failure to lay, 109.

Cardwell, Rt. Hon. Edward, M.P., 167; speech at farewell banquet, 168.

Carlisle, Earl of, 174; speech by, 175, 179.

Carmichael, Sir James, Bart., 49.

Carnarvon, Right Hon. Lord, banquet at Liverpool, 402; *Caroline* s.s., shore-end cable, 1865 expedition, 381.

INDEX

INDEX

INDEX

Glass, Elliot & Co., 131, 283; amalgamated with Gutta-Percha Co,, 376.
Goldsmid, Sir Julian, "Submarine" Co. and government, 79 (note).
Gooch, Mr., C.E. (Sir Daniel Gooch, Bart, M.P.), and the *Great Eastern*, 380; capital for 1866 expedition, 385.
Gorgon, H.M.S., 234; returns to Queenstown, 273,
Gould, Dr., 72, 76, 374.
Grand North Atlantic Telegraph project, 372.
Grapnels, used on 1866 expedition, 388; modern improvements in, 395 (note).
Grappling appliances, first use of, 351; feasibility of in mid ocean, 384.
Graystones (Greystones), old carved mantlepiece at, 13; present occupier of, 13 (note).
Great Eastern : size and failure as cargo vessel, 379; fitted out for second Atlantic cable, 380; recovery of 1865 cable, 394; arrangements for feeding the monster ship with coal, 394 (note).
Great Eastern Co.: represented 1865 expedition, 382; fitness for cable laying work, 384; improved for 1866 expedition, 390,
Greaves, family of, connected with the Brights, 9 (note).
Greenwich, cable works, 381; borough of, 382.
Gray, Earl de, 366.
Griffith, Mr. S., machinery fitting, *Great Eastern*, 380.
Gurney, Mr. Samuel, M.P., 121.
Gutta-percha, for insulating, used by Bright, 54; decay in underground lines, Bright's apparatus to obviate decay, 72; introduced as insulating medium, 95; early cable coverings improved on since, 347; supposed disadvantages of, 347 (note).
Gutta-Percha Co., preliminary contracts with, 131; amalgamated with Messrs. Glass, Elliot & Co., 376.
Gutteres, Mr. James, of the "Magnetic" Co., a survivor of early telegraphy, 76.

Hailstone, *Yorkshire Worthies*, by, 9, 14 (note).
Hallamshire, Hundred of, the, 1; Hunter s work on, 5 (note), 11.
Hamilton, Capt., H.M.S. *Sphinx*, 381.
Hankey, Mr. T. A., 121.
Harrison, Mr. Henry, 48; Anglo-Irish cable, 87.
Harrow Weald, Bright returns to, 333.
Hartnop, Mr. J. C., F.R.A.S., Liverpool Observatory, 65.
Hatcher, Mr. W. H., engineer to Electric Co., 76.

Hawes, Capt., R.N., controls squadron Anglo-Irish cable, 83.
Hemp, use of, in cable armour, 378.
Henley, Mr. W. T., laying of underground wires, 57; magneto-electric instruments, 66; improved magnetic telegraph, 352; manufacture shore-end second Atlantic cable, 381 (note).
Heywood-Jones, Mr. R. H., J.P., present owner of Badsworth Hall, 2 (note).
High Beach, shooting manor of, 22.
Highton, Mr. Edward, 49.
Highton, Rev. Henry, 40.
Hjorth dynamo, Bright's experiments with, 64.
Hoar, Mr., engineer on *Agamemnon*, 300.
Holland-Belgium, early cable between, 92 (note).
Hooghly river, subaqueous telegraph, 94.
Hornby, Mr. T. D., 48.
Horsfall, Mr. T. B., 48.
Hotham, Lord, 2.
Howard, the family of, in Yorkshire, 5 (note).
Hubback, Mr. Joseph, 117.
Hudson, Capt. W. L., U.S.N., 149, 163, 322.
Hughes, Prof. D. E., F.R.S., experiments with type-printing, 203, 358.

Iceland, North Atlantic route to, 360.
Iddesleigh, Earl of. (See under Northcote).
Illustrated London News, articles on *Great Eastern* 1865 expedition, 383.
Illustrious visitors during construction of cable, 139.
"Improvements in apparatus for laying submarine telegraph cables," 152 (note).
India, first line to, 360.
"Indo-European" Telegraph Co., 75.
Institution of Electrical Engineers, Lord Kelvin's address to, 1889, 346 (note).
Insulator, Bright's porcelain, 48; efficiency of, 49; testimony of Lord Kelvin to merit of, 48 (note); terminal, or shackle, 49; when employed, 50; second Atlantic cable, 377.
Insulation, early experiments, 94; tested by Prof. Thomson and Mr. Whitehouse, 198; of conducting wire, system of check on, 290; variation in explained, 295 (note); improvement in methods of, 347.

Jamieson, Prof. Andrew, F.R.S.E., grapnel of, 395 (note).

502

INDEX

INDEX

INDEX

INDEX

Butler & Tanner, The Selwood Printing Works, Frome, and London.

One Volume, Super-Royal Octavo, nearly 800 pages, with numerous Illustrations, Maps, and Diagrams. Price £3 3s. net.

Submarine Telegraphs

Their History, Construction and Working

FOUNDED IN PART ON WÜNSCHENDORFF'S "TRAITÉ DE TÉLÉ-
GRAPHIE SOUS MARINE," AND COMPILED
FROM AUTHORITATIVE AND EXCLUSIVE SOURCES

By CHARLES BRIGHT, F.R.S.E.

Associate Member of the Institution of Civil Engineers
Member of the Institution of Mechanical Engineers
Member of the Institution of Electrical Engineers

EXTRACTS FROM NOTICES OF THE PRESS

" The author deals with his subject from all points of view—political and strategical as well as scientific. The work will be of interest, not only to men of science, but to the general public. We can strongly recommend it." —*Athenæum.*

" This book is full of information. It makes a book of reference which should be in every engineer's library."—*Nature.*

"There are few, if any, persons more fitted to write a treatise on submarine telegraphy than Mr. Charles Bright. The author has done his work admirably, and has written in a way which will appeal as much to the layman as to the engineer. . . . This admirable volume must, for many years to come, hold the position of the English classic on submarine telegraphy. . . . It is a volume which cannot fail to add lustre to the name borne by its author."—*Engineer.*

"Mr. Bright's interestingly written and admirably illustrated book will meet with a welcome reception from cable men."—*Electrician.*

LONDON: CROSBY LOCKWOOD AND SON

7 STATIONERS' HALL COURT, LUDGATE HILL

Milton Keynes UK
Ingram Content Group UK Ltd.
UKHW041520181024
449640UK00009B/102